Modern Grinding Technology and Systems

Modern Grinding Technology and Systems

Special Issue Editor

W. Brian Rowe

MDPI • Basel • Beijing • Wuhan • Barcelona • Belgrade

Special Issue Editor
W. Brian Rowe
Liverpool John Moores University
UK

Editorial Office
MDPI
St. Alban-Anlage 66
4052 Basel, Switzerland

This is a reprint of articles from the Special Issue published online in the open access journal *Inventions* (ISSN 2411-5134) from 2018 to 2019 (available at: http://www.mdpi.com/journal/inventions/special_issues/MGTS)

For citation purposes, cite each article independently as indicated on the article page online and as indicated below:

LastName, A.A.; LastName, B.B.; LastName, C.C. Article Title. *Journal Name* **Year**, *Article Number*, Page Range.

ISBN 978-3-03842-938-8 (HBK)
ISBN 978-3-03842-937-1 (PDF)

Contents

About the Special Issue Editor

W. Brian Rowe, Professor. W. Brian Rowe is mainly known for research related to grinding and grinding machines, including fluid film-bearing technology. His early career centered in the motor and machine tool industries, leading the research and development of new grinding machines. He was awarded a Ph.D. and later a Doctor of Science from Manchester University. He became the Head of Mechanical and Production Engineering at Liverpool Polytechnic, later becoming Assistant Rector of Liverpool John Moores University. Retiring from this post, he became an emeritus professor of the University and Director of the Advanced Manufacturing Technology Research Laboratory, a post he served for more than 10 years. He supervised many doctoral students, to whom he is eternally grateful for their dedication and enthusiasm in working with industrial companies to solve their grinding and process control problems.

Preface to "Modern Grinding Technology and Systems"

Grinding is a precision material removal process employed to improve the shape, size, and surface texture of a wide variety of products. The value of the grinding process has been increasingly recognized because it can achieve high removal rates and very high orders of precision, in many cases surpassing all other manufacturing processes.

Because grinding constitutes the final machining process controlling the quality and surface integrity of manufactured parts, it is often described as a 'strategic' process. It can be considered as the most important stage of manufacturing, being key to achieving high productivity, high accuracy, and high quality.

Many countries around the world have recognized the strategic importance of modern grinding technology and set up specialist research centers to ensure that rapid advances in this field can be made.

This book reviews these issues in Chapter 1, which includes references to wider sources of information on specialist aspects. In this chapter, it is shown how engineering research and engineering ingenuity has led to rapid advances in the productivity and quality of grinding processes. In some cases, slow lapping and polishing processes for very hard materials have been replaced by material removal processes conducted at much higher speeds. The succeeding chapters, written by internationally recognized researchers, illustrate results from investigations of factors that must and should be controlled to achieve optimum grinding performance. The breadth of the subject matter is very wide, ranging across materials analysis, abrasives, wheel designs, machine kinematics, grinding fluids, control systems, and the application of artificial intelligence.

The book will be of interest to industrialists and researchers in the field of manufacturing technology.

W. Brian Rowe
Special Issue Editor

inventions

MDPI

Review

Towards High Productivity in Precision Grinding

W. Brian Rowe

General Engineering Research Institute, Liverpool John Moores University, Liverpool L3 3AF, UK; B.Rowe@ljmu.ac.uk; Tel.: +44-01548-842667

Received: 18 March 2018; Accepted: 10 April 2018; Published: 12 April 2018

Abstract: Over the last century, substantial advances have been made, based on improved understanding of the requirements of grinding processes, machines, control systems, materials, abrasives, wheel preparation, coolants, lubricants, and coolant delivery. This paper reviews a selection of areas in which the application of scientific principles and engineering ingenuity has led to the development of new grinding processes, abrasives, tools, machines, and systems. Topics feature a selection of areas where relationships between scientific principles and new techniques are yielding improved productivity and better quality. These examples point towards further advances that can fruitfully be pursued. Applications in modern grinding technology range from high-precision kinematics for grinding very large lenses and reflectors through to medium size grinding machine processes and further down to grinding very small components used in micro electro-mechanical systems (MEMS) devices. The importance of material issues is emphasized for the range of conventional engineering steels, through to aerospace materials, ceramics, and composites. It is suggested that future advances in productivity will include the wider application of artificial intelligence and robotics to improve precision, process efficiency, and features required to integrate grinding processes into wider manufacturing systems.

Keywords: grinding; processes; wheels; machines; systems; control; removal rates; precision; sensors; micro-grinding; coolant; lubrication; coolant delivery

---

1. What Is the Potential for Future Innovation

In a narrow context, grinding is the removal of material using a high speed abrasive. In a wider context abrasive technology includes material removal using low speed abrasive. Innovations that allow high speed processes to replace low speed processes achieve higher removal rates as explained in more detail by Rowe [1]. For example, the introduction of new grinding techniques such as electrolytic in-process dressing using metal-bonded wheels make it possible to mirror-grind hard silicon wafers with greatly increased removal rates [2].

Examples of low and high speed abrasive processes are illustrated in Figure 1. Higher tool speeds allow much higher removal rates and have therefore been one of the main drivers of technological advance. However, this single predominant trend has required many other advances in machine tool design, grinding wheel design, and system control.

Innovations in modern grinding technology cover an immense range of scientific and industrial disciplines spanning from the interdisciplinary science of tribology and surface interactions to control systems and artificial intelligence [1]. The subject fascinates engineers trying to reduce costs and improve manufacturing productivity within the factory environment and scientists seeking to explain complex interactions between an abrasive and a workpiece. The drivers of innovation include demands for better abrasive tools, higher precision, higher removal rates, introduction of new work materials such as aerospace alloys and hard ceramics, and demands for better surface integrity.

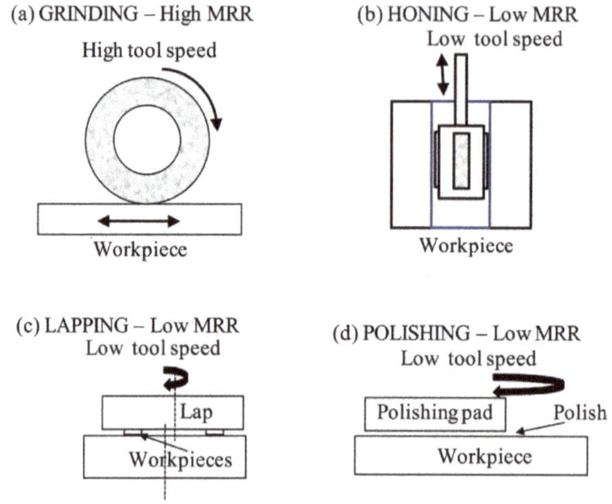

Figure 1. Increased tool speeds has led to greatly increased material removal rates. (**a**) Grinding-High MRR; (**b**) Honing-Low MRR; (**c**) Lapping-LowMRR; (**d**) Polishing-Low MRR

In recent decades, manufacturing engineers are also driven by additional needs such as to operate in a healthier environment, to eliminate toxic waste, to increase manufacturing flexibility, and to employ intelligent control of processes using modern digital technology. The potential for innovation is almost limitless depending ultimately on the engineer's imagination. For example, lead times between processes are reduced by integrating a grinding module into a multi-process and multi-tool system. Manufacturers can also envisage employing a robotic grinding tool together with an automated optical inspection system to remove product defects within an automated machining system. Looking back over the previous century, it is startlingly obvious that the range and scope of technological development has been accelerating, giving great hope for the future.

As manufacturing technology advances, there is increasing awareness that grinding is a "strategic process" in that it is a critical process for the achievement of the best quality of parts and the lowest cost [1,3]. This awareness has led to the formation of specialist grinding research laboratories within a number of universities, particularly in Europe, USA, Japan, and China.

Advances are not always made by a single change of a single element of a grinding system but usually by a combination of changes affecting several elements. The main parameters the engineer has to manage are shown in Figure 2, Chen and Morgan [4]. Controlled variables are those inputs that are open to change either by an engineer, a human operator or by a computer controller. Uncontrolled variables are basically set values not usually continuously controlled, although potentially open to change. As the process operates, there is continuous variation in forces, temperatures, vibration, and grinding power each of which critically affects the output quality of the machined parts and costs.

Parts to be produced vary from very large to very small and from conventional soft engineering steels to very hard steels, alloys, and ceramics. The diverse range of products required to be machined means there is a large range of grinding machines having diverse characteristics. It is not proposed to discuss differences in modern machines and applications in this paper, as it would require too much space for a single paper. However, a schematic of a special-purpose grinding machine is shown in Figure 3 as an example of innovative designs produced to achieve high productivity in the grinding of high precision mirrors and lenses [5]. Whereas most machines are based on an open "C" or "U" shaped structure for convenience of part-loading, a new generation of high precision machines employ a box structure to limit machine deflections. With high-precision drives and control systems, the grinding accuracy achieved was better than 1 μm form accuracy on large 1 m sized parts.

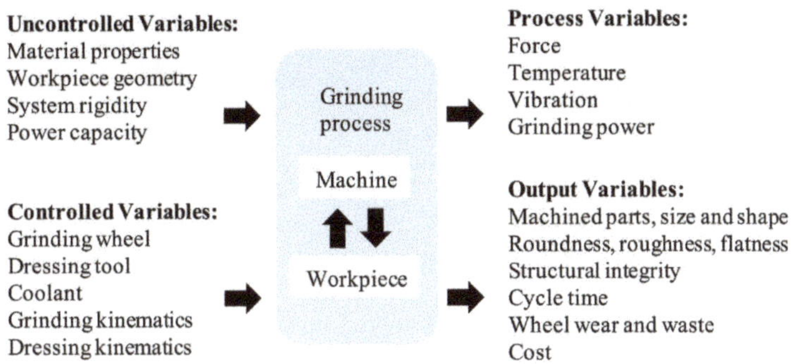

Figure 2. Important elements of a grinding system. Based on Chen & Morgan [4].

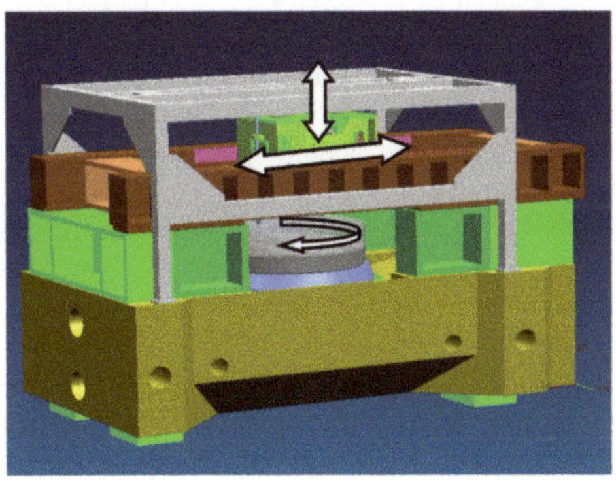

Figure 3. Schematic of Box™ ultra-precision grinder. Shore et al [5].

This paper reviews aspects of science and application leading to improved productivity and quality in grinding processes. Symbols and nomenclature are listed in Appendix A.

2. Wheel Speeds, Material Properties, and Process Kinematics

It was recognized more than 100 years ago that almost every aspect of process behaviour including energy efficiency and material removal rate depends on maximum uncut chip thickness as proposed by Alden [6] and Guest [7]. In this respect, grinding is very little different from micro-milling, although grinding is performed at much high surface speeds of the cutting edges which affects the cutting mechanics. In simple terms, $h_{max} = 2Lv_w a_e / v_s l_c$ where grit spacing $L = \sqrt{3}/Cb_{cu}$ for a sharp triangular grit and $L = \sqrt{3/4}/Cb_{cu}$ for a blunt rectangular grit as shown by Rowe ([1], pp. 319). Increasing chip thickness increases force on the grit. Chip thickness is seen to depend on speeds, feeds, and depth of cut; also on grit spacing and on grit wear as affected by workpiece, coolant and abrasive material properties. The major difference between grinding and micro-milling is the randomness of the grit locations in the wheel surface which means the maximum chip thickness is highly variable in practice. Variation in maximum chip thickness means that not all abrasive grains cut at optimum depth. Many will merely rub on the workpiece wasting energy.

Force on the abrasive grit increases with removal rate and reduces with increasing wheel speed and contact length. However, grinding is also affected by rubbing and ploughing processes. Figure 4 illustrates three different regimes of physical contact that take place with increasing grain penetration [8].

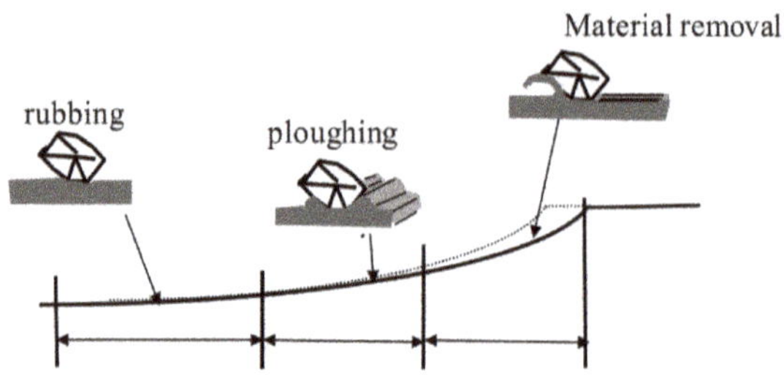

Figure 4. Changes in abrasive regime with increasing grain penetration.

Hahn [8] showed that an abrasive grit rubbing on a ductile material with low normal force increases or reduces surface roughness without removing significant material. This abrasive action is typical of blunt grits. With further increase in normal force, a grit plastically deforms the surface causing side pile-up of material known as ploughing. With yet further increase in force or with a sharper grit, chips are formed and material is removed from the surface. Advances in grinding technology are often associated with changes to reduce wasteful rubbing and ploughing action. Innovations aim to increase the proportion of chip removal compared to ploughing and rubbing so that material is removed with minimal energy.

The transition points between these three regimes depend on various factors notably abrasive sharpness, material properties and lubrication. Processes for polishing and lapping traditionally take place predominantly in the rubbing regime, whereas precision grinding takes place in all three regimes. High energy deep grinding and cut-off operations take place predominantly in the material removal regime by chip formation for ductile materials or by crack formation for brittle materials.

It should be emphasized that energy and material removal in grinding is strongly dependent on the nature of the workpiece material. Hard materials such as cemented carbides [9] and hard ceramics [10] are brittle and difficult to grind. Such materials tend to exhibit a dominance of crack formation when machined but also require a very hard abrasive such as diamond. However, in many cases, it has been shown possible to grind brittle materials in the ductile mode by employing extremely small chip thickness, thus reducing the risks of weakened structures, a process pioneered by Ohmori and Nakagawa [2].

Not all brittle materials are difficult to grind. Some materials, such as grey cast iron, are easy to grind as shown by low energy required compared with grinding hard steels.

Since many parameters influence abrasive contact conditions, there is enormous scope for achievement of higher process efficiency and for achievement of better surface quality. Process efficiency depends primarily on the energy required to remove material. Removal rate per unit width: $Q' = a_e v_w$. Grinding energy per unit unit volume of material removed is known as specific energy: $e = F_t' v_s / a_e v_w$. Figure 5 shows typical relationships between $Q\prime$ and e.

Figure 5a based on Comley et al. [11], shows results for high efficiency deep grinding of camshaft webs at very high removal rates using a sharp electro-plated superabrasive cBN wheel. Figure 5b for surface grinding is based on Malkin [12], and shows results for low speed shallow grinding at 30 m/s using conventional wheels to grind the material AISI 1095HR. Comparing Figure 5a,b, it is seen that huge increases in removal rate of up to 100 times have been achieved in high efficiency deep grinding (HEDG) using high-speed superabrasive wheels together with effective total system design.

Both sets of results show that increasing removal rate by increasing a_e and v_w, reduces specific energy asymptotically. The asymptotic energy level is typically of the order of 10 J/mm^3 but can even be as low as 5 J/mm^3, depending on the work material and abrasive [13].

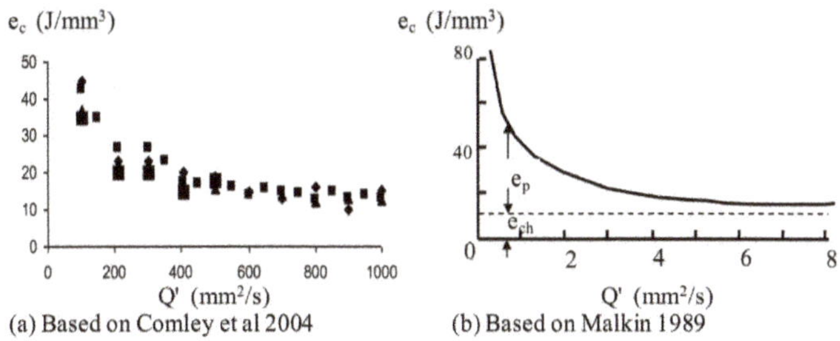

Figure 5. Typical results show specific energy reduces with increasing removal rates. (**a**) Based on Comley et al. [11]; (**b**) Based on Malkin [12].

Malkin [12] concluded that the energy carried away with the chips must be lower than the asymptotic value and that the energy corresponds to the enthalpy for the maximum chip temperature, $e_{ch} = \rho c T_{ch}$ The energy required before steel chips start to become molten, is typically about 6 $\mathrm{J/mm^3}$, approaching a value close to the melting energy of the work material. As melting temperatures are approached, the energy required to shear the material, is reduced. This process and a predominance of chip formation energy over rubbing and sliding energy provides a partial explanation for a low energy asymptote at high removal rates.

Unfortunately, a limit is found to removal rates, when the work material starts to flow into the grinding wheel surface and the abrasive action breaks down. Some materials such as inconel 718, maintain hardness at high temperatures, which makes the material difficult to grind. High forces on the abrasive grits cause rapid wear and high temperatures that increase the tendency for wheel loading. It may be necessary to reduce chip thickness by increasing wheel speed and also to ensure effective lubrication and cooling [14].

3. Process Limits to Removal Rate in Grinding

Increases in removal rate tend to be limited by the need to maintain or improve product quality as specified for accuracy, surface integrity, and surface roughness. Relationships between quality parameters and removal rate depends on many factors such as kinematics, material properties, abrasive properties, machine stiffness, cooling, and lubrication as indicated in Figure 2. Over the years, innovation has allowed removal rates to be increased often with improved quality as a result of increased knowledge, improved understanding of process behaviour and ingenuity employed to expand the process boundaries. Figure 6 shows a limit chart derived for centreless grinding [15].

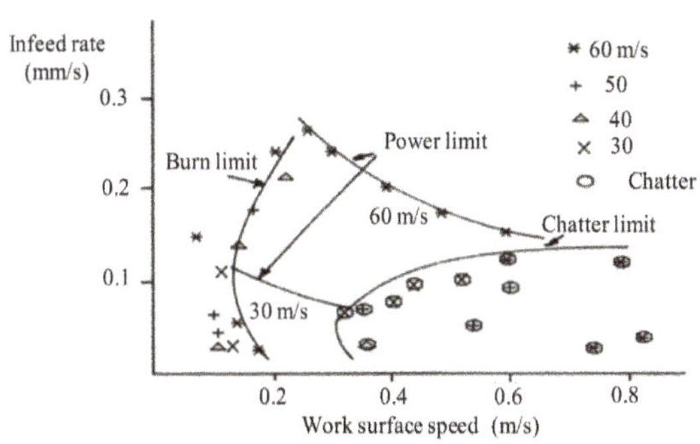

Figure 6. Typical process limit chart for centreless plunge grinding AISI 1055 steel 50 mm dia. × 65 mm long at wheel speeds ranging from 30 m/s to 60 m/s. Based on Rowe et al. [15].

Typical process limits are shown in Figure 6, for infeed rate and work speed where removal rate increases with infeed rate v_f according to $Q' = \pi d_w v_f / 2$ in centreless grinding or $Q' = \pi d_w v_f$ for centre grinding. The safe operating range for the particular grinding wheel, machine design, and work material is enclosed by a thermal damage boundary, the machine power limit, and the chatter limit. This chart is typical for many grinding processes although the nature of the limits may vary. For example, the power limit shown may in many cases be replaced by a maximum surface roughness limit or possibly by a size control limit. The striking feature of the figure is that the maximum infeed rate is greatly increased by increasing the wheel speed from 30 m/s to 60 m/s. This was partly explained by the reduction in chip thickness with increasing wheel speed and corresponding reduction in grinding forces, but was also partly explained by the characteristics of the main spindle speed controller. Much higher grinding speeds, more than 150 m/s have since been made possible by use of superabrasive wheels [16]. It is seen in Figure 6 that low work speeds and high infeed rates lead to burn. It is also found that chatter vibration becomes more of a problem at high work speeds. It can therefore be seen for a particular machine set-up, there is an optimum work speed where a maximum feed rate can be achieved.

4. Process Control for High Productivity and Quality, Use of Sensors, and Process Models

Process control may involve introducing variations in process input parameters to correct for quality errors such as inaccuracy, surface integrity problems due to thermal damage, excessive roughness, and excessive vibration. With repeated batch manufacture of a product, it is possible to apply process control changes between parts produced. Sometimes, with larger workpieces or longer grinding cycles, it may be necessary to apply corrections within a part cycle.

Unfortunately, it is not usually possible to employ data directly from charts such as Figure 6 for process control. This is because the process boundaries shift with time depending on various factors such as grinding wheel wear, workpiece hardness, redressing kinematics, and dressing tool wear. An alternative is to employ process sensors and process models to monitor process changes and to predict necessary changes to bring a process back into a safe operating region.

A major advance in grinding process control in the 20th Century was the introduction of precision diameter gauging linked to CNC control. Typical results for precision grinding with diameter gauging are shown in Figure 7. Typical results for a constant feedrate system are shown in Figure 7a. Employing a 60 s cycle time, it was possible to hold size for successive parts within approximately 1 μm. However, with an adaptive control system, it was possible to predict a new target feed position and to vary the spark-out dwell period for each part [17]. A similar size range was achieved but mean cycle time was reduced from 60 s to 37 s, Figure 7b. With adaptive control, removal rate was substantially improved without sacrificing accuracy.

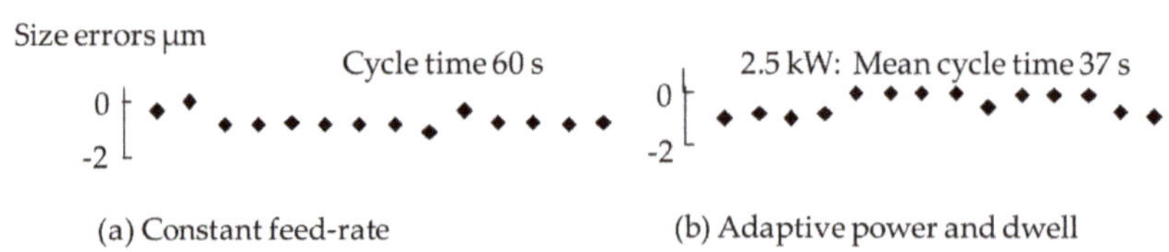

Figure 7. Size variations using diameter gauging with position control: (**a**) For constant feed-rate and (**b**) for adaptive power and dwell. Based on Liverton & Rowe [17].

Using adaptive control, it is necessary to use sensors that can detect whether the system is being maintained within an acceptable operating region while increasing the control variable towards the process limit. Various sensors have been used, in addition to the standard position sensors employed for CNC, the most common being a power sensor which is reasonably inexpensive and easy to incorporate into a system without requiring a dynamometer fixture. Other sensors, including size sensors, force sensors, acoustic emission sensors, and Barkhausen noise sensors may be employed to check for safe operation. To overcome many problems, such as burn or chatter, it may be necessary to reduce work speed or to reduce infeed rate or possibly to redress the grinding wheel. Some decisions require an intelligent input either from a human operator or possibly from AI software incorporated into the control system. For process control, it is often necessary to employ a process model to relate a measured variable to the required control variable. For example, to adaptively control spark-out dwell period, as in Figure 7b, the best way was to determine system time constant from the measured power [18].

For prevention of thermal damage in industrial processes, a power sensor is usually employed, either in conjuction with a thermal model or without a thermal model [1,3]. Of course, there are other decisions that affect the onset of thermal damage such as redressing a worn wheel or even replacing the grinding wheel with a more efficient wheel.

The onset of thermal damage correlates with an excessive grinding power for the particular grinding conditions or an excessive force level. In Figure 8, using a force sensor, feed rate is increased in small steps up to the limit for thermal damage, Steffan et al. [19]. The application of post-process Barkhausen noise measurement for correlation with temperature modeling is described by Sridharan et al. [20]. This latter paper introduces the concept of modelling and correlating more than one measured variable for improved reliability of prediction. The use of multiple sensors will no doubt be a trend for the future.

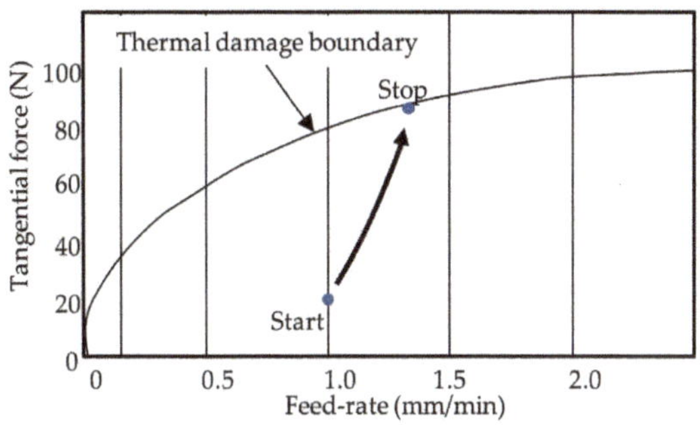

Figure 8. Adaptive control of feed-rate avoiding thermal damage. Based on Steffan et al. [19].

5. Temperature Rise Modeling

Thermal models may be used as a basis for process control, Rowe [21]. Temperature rise is critical for workpiece surface hardness, surface roughness, material phase changes, and onset of tensile residual stress. The Rowe thermal model has been validated for a range of grinding processes including conventional shallow grinding, creep feed grinding, and HEDG [21].

The model can be derived from heat balance of the heat flows in the contact zone. Four heat flows make up the total heat flow, Werner et al. [22]. The total heat flow

results from the grinding power P dissipated within the grinding contact area $l_c b_w$ and is defined as $q \equiv P/l_c b_w$. The four heat flows take place to the workpiece, the abrasive grains, the chips and the grinding fluid. The chip energy q_{ch}, see Section 2, is immediately carried away from the contact area and plays no further part in heat partitioning. The remaining heat $q - q_{ch}$, is shared at the grain and workpiece rubbing interface according to Hahn [8] and also according to the more refined model of Rowe et al. [21,23]. The heat flow into the workpiece at the grain contact is given by $R_{ws}(q - q_{ch})$. Some heat q_f is conducted out again by the fluid within the grinding contact area. The temperature rise, of the workpiece is therefore given by the Rowe thermal model [1,21], as follows:

$$\Delta T = R_{ws}(q - q_{ch}) / \left(h_w + h_f\right)$$

where, R_{ws} has a value between 0 and 1 depending on the grain conductivity, the workpiece thermal properties, the grit contact area, and the wheel speed. For conventional abrasives when dry grinding steels, a typical value is 0.85, and for superabrasive cBN, a typical value is 0.46. For diamond abrasive a typical value is even lower, showing a great advantage of superabrasives for avoiding large temperature rise.

The terms h_w and h_f are convection factors for the workpiece and the fluid respectively. The convection factor for the workpiece is given by classical heat transfer theory and defined as $h_w \equiv q_w/\Delta T$ given by,

$$h_w = \frac{\beta_w}{C} \cdot \sqrt{\frac{v_w}{l_c}}$$

where $C \approx 1$ for conventional speeds. For shallow cuts and deep cuts and other values of speed, values of C are given in Figure 9 and in the literature [1,21]. Temperatures on the finish surface are seen to be lower than on the grinding contact surface for deep grinding.

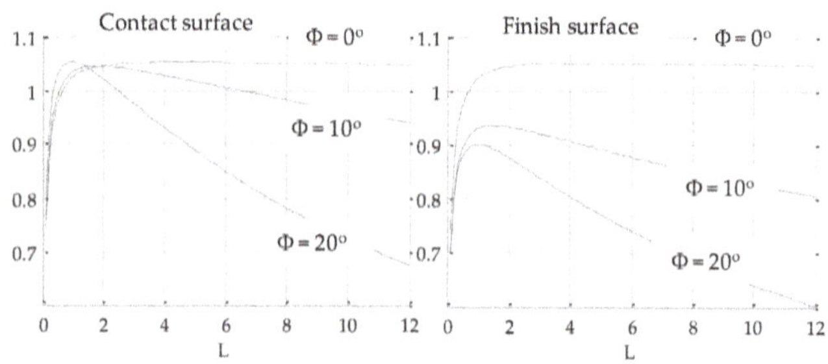

Figure 9. C-factors for temperature rise where Peclet Number $L = v_w l_c / 4\alpha$.

It can be seen that a process model for fluid film convection is highly desirable to form a link between grinding energy and workpiece temperature rise. This is still a subject of research. An expression sometimes used to estimate fluid convection assumes the pores of the wheel are filled with cold fluid throughout the contact arc and cools the workpiece as though it was a "solid-fluid wheel". The convection factor from this assumption is given by $h_f = (\beta_f/C).\sqrt{v_s/l_c}$. Correlation with experimental results, Figure 10 shows reasonable agreement in low temperature grinding below the fluid boiling temperature [24]. It is also found that fluid convection can be much higher than previously realized. In practice, the convection factor rapidly reduces towards zero if the fluid boils.

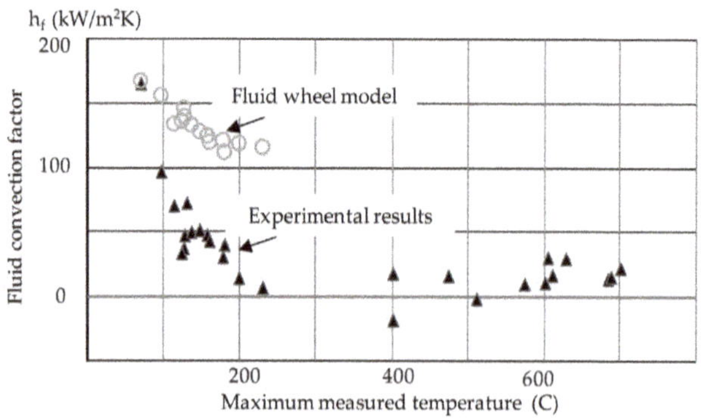

Figure 10. Model for water based fluid agrees best for grinding temperature below 100 C. Data from [24].

Simplifications, for example, ignoring fluid convection, conduction into the wheel, and chip energy, over-estimate temperature rise and grossly over-estimate temperature rise in some cases to the point that steel would apparently reach 10,000 °C or more. A simplified but inaccurate assumption that all heat enters the workpiece, gives the misleading expression $\Delta T = q\sqrt{l_c}/\beta_w\sqrt{v_w}$.

6. Grinding Fluids and Fluid Delivery

Grinding fluids play several important roles in a liquid-based wet grinding process. As seen in the previous section, a grinding fluid can play a valuable role in reducing grinding temperature by providing surface cooling within the grinding contact area. A major early review of friction, cooling, and lubricaton in grinding lists 160 research papers (Brinksmeier et al. [25]). Other roles are played by the grinding fluid including keeping the machine cool, keeping the body of the workpiece cool and flushing away the grinding debris. This latter role is particularly important in high removal rate grinding and in mirror finish grinding. Grinding swarf can clog a grinding wheel or be pulled into the grinding contact area causing surface damage [26]. The grinding fluid also serves a physical–chemical lubrication role in the contact between the work material and the abrasive material which can be critical for avoidance of excessive grinding wheel wear and rapid onset of wheel loading [27].

There are three primary classes of grinding fluids: (1) fluids which are mixed with water known as water-based fluids and emulsions; (2) mineral oils; and (3) synthetic oils. Within these groups there are a very diverse family of oils and chemicals selected according to the type of abrasive and the nature of the work material.

Water-based coolants have a better cooling property than oils but boil at a much lower temperature. The much reduced cooling above the boiling temperature of water based coolants is seen in Figure 9. Water-based coolants also need to be changed frequently before excessive contamination and bio-degradation take place. Neat oil coolants have better chemical stability and are generally better for preventing subsequent corrosion of the workpieces.

Coarser grit wheels allow more fluid to circulate into the grinding contact and are generally better at maintaining the cutting efficiency of the wheel [28]. The supply of fluid into the grinding zone is impeded however by the boundary layer that surrounds a speeding grinding wheel [29]. The air barrier can be overcome by providing a fluid jet velocity at a similar speed to the wheel speed and an ample supply of fluid [30]. However, it is wasteful to supply too much fluid as the grinding surface is unable to accommodate a large excess in the pores. This can be seen from Figure 11. If the fluid supply is adequate, the quantity of fluid that actually enters the grinding contact increases with wheel speed. However, when the wheel speed is too high, the supply of fluid is no longer sufficient so that the useful flow levels off and will actually start to reduce. The shape and profile of the nozzle affects the effectiveness of delivery. Round nozzles are generally better than rectangular nozzles but a single round nozzle cannot cope with wide wheels. In either case, an internal concavely converging feed to the nozzle is better than a convexly converging nozzle [31].

It is possible to roughly optimize the velocity and the flowrate required to maximize the useful flow that enters the grinding contact quite simply based on the wheel porosity, taken from the wheel manufacturer's data, the wheel speed and the wheel width [32]. Of course, there are times when much smaller quantities of fluid should be applied, as when finishing a product to achieve a very fine size dimensional tolerance. The fluid causes a hydrodynamic force that pushes the grinding wheel away from the ground surface. Under these circumstances, it is necessary to reduce the removal rate and reduce the flowrate to limit the hydrodynamic force.

There is a cost of fluid cooling both financial and environmental [33]. Fluid supply requires energy to provide the high velocities and supply rates needed for high removal rate grinding at high wheel speeds. Unless the jet velocity is sufficient, the fluid jet causes a drag on the grinding wheel because the fluid has to be accelerated up to grinding wheel speed. The power required taken through the wheel drive motor can be substantial, particularly when using a shoe nozzle.

Environmental concerns have led to the introduction of minimum quantity lubrication (MQL) for grinding [34]. MQL is usually more appropriate for low removal rate grinding since the cooling achieved is much lower than with liquid coolants. Conventional cooling delivery for grinding may involve many litres/minute whereas the delivery for MQL is usually in the form of a fine mist of esters or mineral oils injected into an air stream involving millilitres/min. Good results have been achieved for some finish grinding operations.

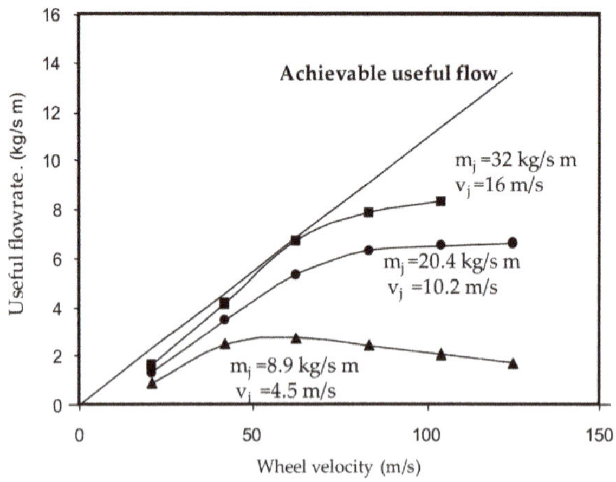

Figure 11. Useful flowrate for a high-porosity grinding wheel. Based on Gviniashvili [30].

7. Grinding Wheels and Abrasives

Conventional grinding wheels are largely based on vitrified bond aluminium oxide abrasives or on vitrified bond silicon carbide abrasives, although other abrasives are employed for specialist purposes. For example, for flexible wheels, resin, rubber or polymer bonds are often employed. Most conventional vitrified grinding wheels run at speeds of 20–45 m/s although with rigorous attention to safety requirements, vitrified wheels can be obtained for higher speeds. Monolayer superabrasive grinding wheels, such as diamond and cBN wheels often run at much higher speeds, in excess of 150 m/s. Although, superabrasives are more expensive, much greater hardness allows these materials to achieve much longer life when grinding very hard mechanical engineering and electronic materials.

In more recent decades, there have been substantial developments in abrasive technology and wheel designs allowing very high speeds to be used as detailed by the late Mike Hitchiner formerly of Saint Gobain [16]. Examples of conventional silicon carbide and alumina grains are shown in Figure 12a,b, courtesy of Saint Gobain. Whereas conventional grains tend to be more blocky in shape, a new generation of more advanced sintered grain abrasives, as shown in Figure 12c,d, have very large length to width ratio. This allows the grains to present a sharper profile when grinding and also allows harder grains to be employed to yield longer wheel life. Large length to width ratio of the grains accommodates more fluid in the pores for porosity and better cooling. The manufacturers tend to refer to such structures as superabrasive ceramics. The new structures have proven very successful in high removal rate grinding with harder materials that quickly blunt more conventional abrasives.

Figure 12e,f shows examples of brazed and electro-plated monolayer superabrasive CBN wheel structures. Monolayer wheels allow very high wheel speeds [35]. Multilayer vitrified CBN wheels are also available as a replacement for conventional wheels. CBN is not only harder than conventional abrasives but has excellent heat conduction properties resulting in lower grinding temperatures [1,36].

High speed vitrified wheels requires extreme care to design for the stresses involved and avoid wheel failures [37]. This is because the vitrified bond is the source of potential failure whereas monolayer and multilayer metal-bond wheels provide the capability to run at the highest speeds.

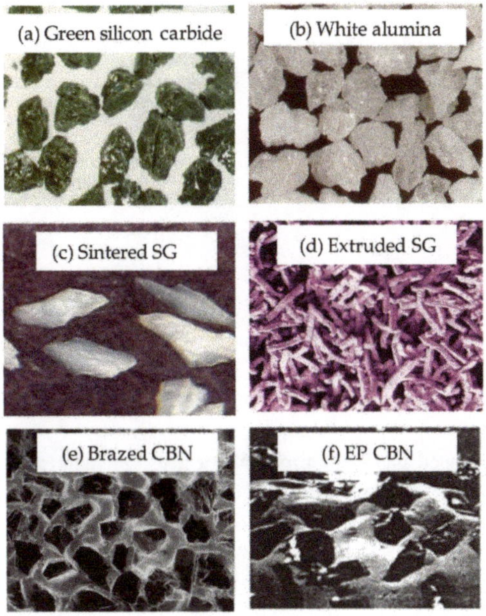

Figure 12. Conventional abrasives, superabrasive SG, and CBN. Courtesy of Saint Gobain Abrasives. (**a**) Green silicon carbide; (**b**) white alumina; (**c**) sintered SG; (**d**) extruded SG; (**e**) Brazed CBN; (**f**) EP CBN.

Diamond is harder than CBN, but is generally unsuited to grinding steels because carbon diffuses from the diamond causing rapid wheel wear. However, diamond due to its hardness has advantages for very hard materials.

Conventional vitrified bond wheels are generally dressed using diamond tools to correct the profile accuracy and to sharpen the wheel surface. Dressing time is non-productive so it is an advantage if wheels can grind for a long time before needing to be redressed [38]. Vitrified CBN wheels may also be dressed but dressing consumes valuable CBN material. A technique to minimize the material removed in dressing and to optimize subsequent grinding performance is known as 'touch dressing'. Acoustic emission sensors detect the instant of contact between the dressing tool and the wheel, hence allowing accurate setting of a minimal dressed depth [39].

Monolayer CBN and monolayer diamond wheels are not usually dressed since this would damage the abrasive layer. Such wheels must therefore be manufactured with great accuracy to ensure the grains all lay at the same level on the surface. Manufacturers aim to control the evenness and circularity of the wheel to 1 micron.

Some modern dense materials such as hard ceramics require diamond or CBN grinding because of their hardness and brittle characteristics. It is often possible to grind such materials and obtain mirror surface finishes using extremely small depth of cut. The depth of cut has to be small enough that material removal takes place primarily in a ductile mode. Large depths of cut lead to brittle mode deformation causing surface cracks and probable failure. Processes developed in recent decades involve the use of fine-grained CBN or diamond in metal-bond wheels. Dressing such wheels has been developed using electro-chemical and electrolytic removal processes. For example, a particular process known as in-process electrolytic dressing, (ELID) is employed in a grinding process to mirror-grind silicon wafers [2,40]. Electrolytic dressing exposes the abrasive grains and thus allows a grinding process to take place. Grinding replaces a lapping process with much improved productivity and accuracy. Basic elements of an ELID system are shown in Figure 13.

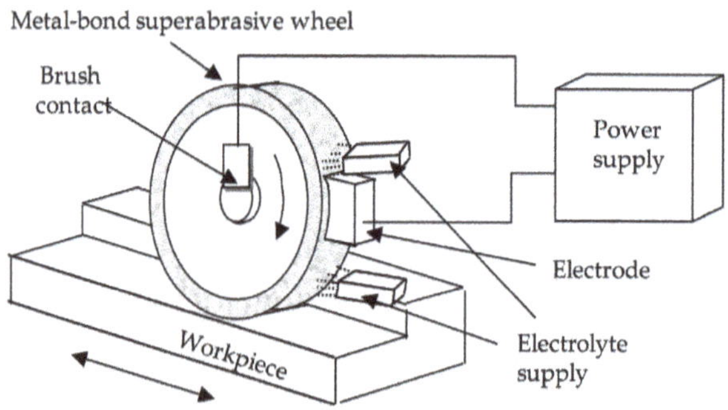

Figure 13. Electrolytic in-process dressing system.

In recent decades, a need has been apparent for grinding wheels that can grind micro-tools which may have a tip diameter of a few tens of micrometers, Ohmori et al. [41]. Such tools have been successfully ground using the ELID process. Aurich et al., also describe the machining of micro-parts [42]. Figure 14 shows relatively large micro-tools produced by Butler-Smith et al. [43]. Figure 14a shows a tool where fine diamond grains are secured on the surface by electro-plating and Figure 14b shows a tool where the regular pattern of cutting edges was produced by laser ablation of solid diamond formed by chemical vapour deposition (CVD). The regularity and even spacing of the cutting edges of the laser ablated tool was said to give improved surface roughness and accuracy.

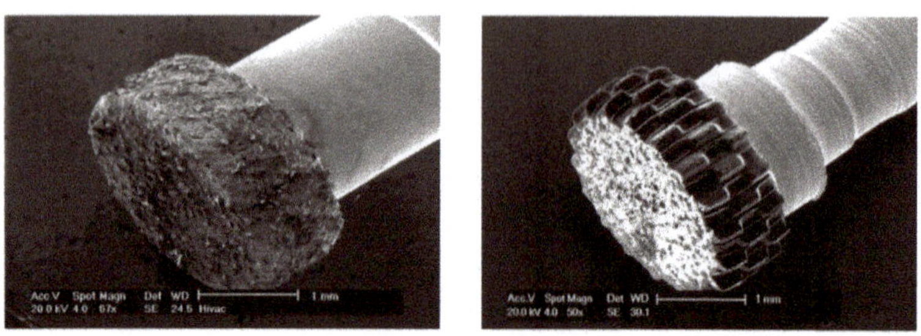

(a) (b)

Figure 14. Micro-grinding wheels: (**a**) electro-plated diamond wheel; (**b**) laser-ablated diamond wheel. Photographs supplied and reproduced by permission of P W Butler-Smith.

The recent rapid increase in parts produced by additive manufacturing has created a need for finishing processes that can improve the surface roughness of parts having complex geometrical shapes. An alternative to grinding using a grinding wheel is the use of abrasive mass flow processes as described by Jamal and Morgan [44].

8. Application of Artificial Intelligence and System Integration

Modern CNC systems contain intelligent software for various functions such as position, velocity, and acceleration control. AI generally implies decision-making that is, to some extent, judgemental based on sensor input. It implies that the machine can make a decision that otherwise may require a human being. Whereas a human being may become tired, the AI machine will continuously perform repetitive tasks without human intervention, making adjustments where necessary. AI may be used to help control the grinding process by reacting to sensors within the grinding machine and measurement equipment as mentioned in Section 4. AI analyses and translates data from measurements to make decisions such as to increase or reduce feedrate to avoid thermal damage or to initiate a change point in a feed cycle as described above or to raise an alarm call for operator intervention. The application of standard AI tools such as artificial neural networks, genetic algorithms, and case-based reasoning in grinding systems are further discussed in a review paper [45].

With the passage of time it becomes clear, that the ability of computers to collect, process, and analyse data is becoming ubiquitous. Data can be collected and analyzed for control and scheduling of manufacture in a wider sense. For the user and indeed for the machine tool manufacturer there is the possibility to collect data concerning the need for maintenance of the grinding machine or the effectiveness of its operation, using the Internet of Things (IoT). A further way in which AI can help is to assist in selecting appropriate tools or grinding conditions for the initial process set-up. An example is the selection of an appropriate grinding wheel including wheel grade and structure using case-based reasoning [46].

9. Conclusions

This paper reviews a selection of developments in which the application of science and ingenuity has led to new grinding processes, abrasives, tools, machines, and systems. It is shown that improved technology has yielded higher productivity and better quality across a diverse range of applications involving very large and very small products. Modern grinding technology has learned to cope with a diverse range of materials relying on new abrasive tools and various kinematic operating conditions. It is concluded that this trend will continue and that future advances in productivity will include the wider application of artificial intelligence and robotics to improve precision, process efficiency, and features required to integrate grinding processes into wider manufacturing systems and into versatile machine tools.

Appendix A. Notation

a_e	Real depth of cut
b_{cu}	Uncut grit contact width
b_w	Width of grinding contact
c	Specific heat capacity of work material
C	Number of active grits per unit area
d_e	Equivalent wheel diameter
e or u	Specific energy to remove unit volume of material
h_m	Maximum uncut chip thickness
v_s	Wheel surface speed
k	Thermal conductivity
l_c	Grinding contact length
L	Effective tangential grit spacing or
L	Peclet number
P	Grinding power
q	Heat flux given as heat flow per unit contact area
q_{ch}	Heat flux to chips
q_w	Heat flux to workpiece within contact area
q_s	Heat flux to wheel within contact area
q_f	Heat flux to grinding fluid within contact area
Q'	Removal rate per unit width

ΔT	Temperature rise
T_{ch}	Temperature rise of chips
v_f	Normal feed or infeed rate
v_s	Wheel surface speed
v_w	Workpiece surface speed
α	Thermal diffusivity of work material
β	Thermal property $\sqrt{k\rho c}$
ρ	Mass density of work material
$\varnothing = \mathbf{l_c}/\mathbf{d_e}$	Grinding contact angle

References

1. Rowe, W.B. *Modern Grinding Technology*, 2nd ed.; Elsevier, William Andrew Imprint: Waltham, MA, USA; Oxford, UK, 2014.
2. Ohmori, H.; Nakagawa, T. Mirror Surface Grinding of Silicon Wafers with Electrolytic in Process Dressing. *CIRP Ann.* **1990**, *39*, 329–332. [CrossRef]
3. Inasaki, I.; Toenshoff, H.K.; Howes, T.D. Abrasive Machining in the Future. *CIRP Ann.* **1993**, *42*, 723–732. [CrossRef]
4. Chen, X.; Morgan, M.N. Advances in Quality and Productivity in Precision Grinding—A Review of Selected Research. In Proceedings of the ASME International Manufacturing Science & Engineering Conference, Symposium on Abrasive Machining, Blacksburg, VA, USA, 27 June–1 July 2016.
5. Shore, P.; Morantz, P.; Luo, X.; Tonnelier, X.; Collins, R.; Roberts, A.; May-Miller, R.; Read, R. Big OptiX Ultra-Precision Grinding/Measuring System. *Proc. SPIE* **2005**, *5965*, 241–248.
6. Alden, G.L. Operation of Grinding Wheels in Machine Grinding. *ASME Trans.* **1914**, *36*, 451–460.
7. Guest, J.J. *Grinding Machinery*; Edward Arnold: London, UK, 1915.
8. Hahn, R.S. On the Nature of the Grinding Process. In Proceedings of the 3rd Machine Tool Design and Research Conference, Birmingham University, Birmingham, UK, September 1962; Pergamon Press: Oxford, UK, 1962; pp. 129–154.
9. Wirtz, C.; Mueller, S.; Mattfeld, P.; Klocke, F. A Discussion on Material Removal Mechanisms of Cemented Carbides. *J. Manuf. Sci. Eng.* **2017**, *139*, 121002. [CrossRef]
10. Marinescu, I.D.; Toenshoff, H.K.; Inasaki, I. (Eds.) *Handbook of Ceramic Grinding and Polishing*; Noyes Publications: Park Ridge, NJ, USA; William Andrew Publishing: Norwich, NY, USA, 2000.
11. Comley, P.; Stephenson, D.J.; Corbett, J. High Efficiency Deep Grinding and the Effect on Surface Integrity. *Key Eng. Mater.* **2004**, *257–258*, 207–212. [CrossRef]
12. Malkin, S. *Grinding Technology*; Ellis Horwood: Chichester, UK, 1989.
13. Rowe, W.B.; Jin, T. Temperatures in High Efficiency Deep Grinding. *CIRP Ann.* **2001**, *50*, 205–208. [CrossRef]
14. Rowe, W.B.; Ebbrell, S. Process Requirements for Cost-Effective Precision Grinding. *CIRP Ann.* **2004**, *53*, 255–258. [CrossRef]
15. Rowe, W.B.; Bell, W.F.; Brough, D. Optimization Studies in High Removal Rate Centreless Grinding. *CIRP Ann.* **1986**, *35*, 235–238. [CrossRef]
16. Marinescu, I.D.; Hitchiner, M.P.; Uhlmann, E.; Rowe, W.B.; Inasaki, I. *Handbook of Machining with Grinding Wheels*, 2nd ed.; CRC Press, Taylor and Francis: Boka Raton, FL, USA, 2016.
17. Liverton, J.; Rowe, W.B. Adaptive Control of Cylindrical Grinding—From Development to Commercialization. In Proceedings of the 5th International Society of Manufacturing Engineers Grinding Conference, Cincinnati, OH, USA, 26–28 October 1993.
18. Thomas, D.A.; Allanson, D.R.; Moruzzi, J.L.; Rowe, W.B. In-process Identification of Time Constant for the Control of Grinding. *J. Eng. Ind.* **1995**, *117*, 194–201. [CrossRef]
19. Steffan, M.; Haas, F.; Pierer, A.; Jens, G. Adaptive Grinding Process—Prevention of Thermal Damage Using OPC-UA Technique and In Situ Metrology. *J. Manuf. Sci. Eng.* **2017**, *139*, 121008. [CrossRef]
20. Sridharan, U.; Bedekar, V.; Kolarits, F. A Functional Approach to Integrating Temperature Modeling and Barkhausen Noise Analysis for Prediction of Surface Integrity in Bearing Steels. *CIRP Ann.* **2017**, *66*, 333–336. [CrossRef]
21. Rowe, W.B. Temperatures in Grinding—A Review. *J. Manuf. Sci. Eng.* **2017**, *139*, 121001. [CrossRef]

22. Werner, P.G.; Younis, M.A.; Schlingensiepen, R. Creep-feed—An Effective Method to Reduce Workpiece Surface Temperatures in High Efficiency Grinding Processes. In Proceedings of the 8th North American Metalworking Research Conference, Rolla, MO, USA, 18–21 May 1980; Soc. of Manuf. Engineers; pp. 312–319.
23. Rowe, W.B.; Black, S.; Mills, B.; Qi, H. Analysis of Grinding Temperatures by Energy Partitioning. *Proc. Inst. Mech. Eng. Part B* **1996**, *210*, 579–588. [CrossRef]
24. Zhang, L.; Rowe, W.B.; Morgan, M.N. An Improved Fluid Convection Solution in Conventional Grinding. *Proc. Inst. Mech. Eng. Part B* **2013**, *227*, 832–838. [CrossRef]
25. Brinksmeier, E.; Heinzel, C.; Wittmann, M. Friction, Cooling and Lubrication in Grinding. *CIRP Ann.* **1999**, *48*, 581–598. [CrossRef]
26. Heinzel, C.; Antsupov, G. Prevention of Wheel Clogging in Creep-feed Grinding by Efficient Tool Cleaning. *CIRP Ann.* **2012**, *61*, 323–326. [CrossRef]
27. Marinescu, I.D.; Rowe, W.B.; Dimitrov, B.; Ohmori, H. *Tribology of Abrasive Machining Processes;* William Andrew: Waltham, MA, USA; Oxford, UK, 2013.
28. Engineeer, F.; Guo, C.; Malkin, S. Experimental Measurement of Fluid Flow through the Grinding Zone. *J. Eng. Ind.* **1992**, *114*, 61–66. [CrossRef]
29. Ebbrell, S.; Woolley, N.H.; Tridimas, Y.D.; Allanson, D.R.; Rowe, W.B. The Effects of Cutting Fluid Application on the Grinding Process. *Int. J. Mach. Tools Manuf.* **2000**, *40*, 209–223. [CrossRef]
30. Gviniashvili, V.; Webster, J.; Rowe, W.B. Fluid Flow and Pressure in the Grinding Wheel Workpiece-Interface. *J. Manuf. Sci. Eng.* **2005**, *127*, 201–205. [CrossRef]
31. Rouly, E.; Bauer, R.J.; Warkentin, A. An investigation into the effect of nozzle shape and jet pressure in profile creepfeed grinding. *Proc. Inst. Mech. Eng. Part B. J. Eng. Manuf.* **2017**, *231*, 1116–1130. [CrossRef]
32. Morgan, M.N.; Jackson, A.R.; Baines-Jones, V.; Batako, A.D.; Rowe, W.B. Optimization of Fluid Application in Grinding. *CIRP Ann.* **2008**, *57*, 363–366. [CrossRef]
33. Howes, T.D.; Toenshoff, H.K.; Heuer, W. Environmental Aspects of Grinding Fluids. *CIRP Ann.* **1991**, *40*, 623–630. [CrossRef]
34. Tawakoli, T.; Hada, M.J.; Sadeghi, M.H.; Daneshi, A.; Stockart, S.; Rasifard, A. An Experimental Investigation of the Effects of Workpiece Grinding Parameters in MQL. *Int. J. Mach. Tools Manuf.* **2009**, *49*, 924–932. [CrossRef]
35. Webster, J.; Tricard, M. Innovations in Abrasive Products for Precision Grinding. *CIRP Ann.* **2004**, *53*, 597–642. [CrossRef]
36. Hitchiner, M.P.; McSpadden, S. Evaluation of Factors Controlling CBN Abrasive Selection for Vitrified Bonded Wheels. *Key Eng. Mater.* **2004**, *257–258*, 267–272. [CrossRef]
37. Barlow, N.; Rowe, W.B. Discussion of Stresses in Plain and Reinforced Cylindrical Grinding Wheels. *Int. J. Mach. Tool Des. Res.* **1983**, *23*, 153–160. [CrossRef]
38. Rowe, W.B.; Chen, X. The Identification of Dressing Strategies for Optimal Grinding Performance. In Proceedings of the Thirtieth International MATADOR Conference, Manchester, UK, 31 March–1 April 1993; pp. 195–202.
39. Rowe, W.B.; Chen, X.; Allanson, D.R. The Coolant Coupling Method Applied to Touch Dressing in High Frequency Internal Grinding. In Proceedings of the Thirty-Second International Matador Conference, Manchester, UK, July 1997; UMIST, Macmillan Press: Oxford, UK, 1997; pp. 337–340.
40. Lui, J.H.; Pei, Z.J.; Fisher, G.R. ELID Grinding of Silicon Wafers: A Literature Review. *Int. J. Mach. Tools Manuf.* **2007**, *47*, 529–536.
41. Ohmori, H.; Katahira, K.; Narusa, T.; Uehara, Y.; Nakao, A.; Mizutani, M. Microscopic Effects on Fabrication of Ultra-Fine Micro Tools. *CIRP Ann.* **2007**, *56*, 569–572. [CrossRef]
42. Aurich, J.C.; Engmann, J.; Schueler, G.M.; Haberland, R. Micro-Grinding Tool for Manufacture of Complex Structures in Brittle Materials. *CIRP Ann.* **2009**, *58*, 311–314. [CrossRef]
43. Butler-Smith, P.W.; Axinte, D.A.; Daine, M. Solid Diamon Tools: From Innovative Design and Fabriaction to Preliminary Performance Evaluation in Ti–6Al–4V. *Int. J. Mach. Tools Manuf.* **2012**, *59*, 55–64. [CrossRef]
44. Jamal, M.; Morgan, M.N. Design Process Control for Improved Surface Finish of Metal Additive Manufactured Parts of Complex Build Geometry. *Inventions* **2017**, *2*, 36. [CrossRef]

45. Rowe, W.B.; Li, Y.; Inasaki, I.; Malkin, S. Applications of Artifical Intelligence in Grinding. *CIRP Ann.* **1994**, *43*, 521–532. [CrossRef]
46. Li, Y.; Mills, B.; Rowe, W.B. An Intelligent System for Selection of Grinding Wheels. *Proc. Inst. Mech. Eng. Part B J. Eng. Manuf.* **1997**, *211*, 635–641. [CrossRef]

inventions

MDPI

Article

Modeling and Analysis of Contact Conditions during NC-Form Grinding of Cutting Edges

Eckart Uhlmann and Joachim Bruckhoff *

Institute for Machine Tools and Factory Management, Technical University Berlin, 10623 Berlin, Germany; uhlmann@iwf.tu-berlin.de

* Correspondence: bruckhoff@iwf.tu-berlin.de; Tel.: +49-30-314-23473

Received: 31 May 2017; Accepted: 29 June 2017; Published: 5 July 2017

Abstract: Due to increasing demands on cutting tools, cutting edge preparation is of high priority because of its influence on the tool life. Current cutting edge preparation processes are mostly limited to generating simple roundings on the cutting edge. Multi-axis high precision form grinding processes offer great potential to generate defined cutting edge microgeometries. Knowledge about the relation between grinding strategy and material removal rate can achieve improved work results with regard to higher precision of shape and dimensional accuracy as well as enhanced cutting edge quality. Therefore, a kinematic-geometric model was developed in order to analyze the complex contact conditions during grinding cutting edge microgeometries by using a simulation approach based on the intersection of geometric bodies. The subsequent grinding tests largely validated the utilized simulation approach.

Keywords: NC-form grinding; cutting edge; contact conditions

1. Introduction

Increasing demands on cutting tools with regard to quality and economy have led to constant progress in tool development and tool production. The cutting edge in particular is attracting much interest as it is an essential tool element for the cutting process. The properties of the cutting edge influence the process characteristics and the work results in machining with defined cutting edges [1,2]. The properties of the cutting edge can be divided into physical and chemical characteristics as well as into the geometry and surface topography of the cutting edge. The macro and micro geometrical design of the cutting edge affects the chip formation and the chip removal [3,4]. Investigations focusing on the effect of the cutting edge shape on tool wear in turning experiments showed that cutting edges rounded by brushing have a higher stability and therefore do not tend to have cutting edge breakouts [5]. Further studies about different methods for cutting edge preparation of micro mills made of tungsten carbide showed that magnetic abrasive finishing and drag finishing result in favorable wear behavior because of homogeneous cutting edge profiles [6]. The most common production processes for cutting edge preparation are magnetic abrasive finishing, abrasive blasting, laser machining, brushing, abrasive flow machining, vibratory finishing and drag finishing. These production processes usually generate simple cutting edge roundings. Until now, systematic studies on the influence of different defined cutting edge microgeometries were not possible, as all common cutting edge preparation processes are not able to produce complex and exact defined forms. Developing manufacturing processes to produce defined and complex cutting geometries in the industrial and scientific environment can increase the performance of cutting processes. Multi-axis high precision form grinding processes offer great potential to generate defined cutting edge microgeometries.

Due to its positive hardness and toughness characteristics, cemented carbide is often used as a material in cutting tools [7]. Most cemented carbides consist of a cobalt matrix and the hard material tungsten carbide. The cutting material is produced in a sinter process. The grain size of the sintered powder, as well as the mixing ratio, influences the properties of the material. Due to the high hardness of tungsten carbide of approximately Vicker Hardness 2000 HV0.05, economical machining of cemented carbide can be achieved by grinding processes only [8].

NC (Numerical Control)-form grinding processes with mounted points are used for manufacturing-free forms on difficult-to-machine materials. Examples of these are the machining of optical glasses and ceramic implants or prostheses for dental and medical applications. The main challenges in NC-form grinding are the positional accuracy and the stiffness of the machine axes due to the simultaneous axis movement. Furthermore, the complex tool paths mostly require NC-programming with CAD/CAM (Computer-aided Design/Computer-aided Manufacturing)technology. Tool paths are graphically planned on CAD models and converted into an appropriate NC-Code by post-processors.

Different form grinding strategies lead to different courses of chip cross-sectional areas and thus different material removal rates. As a result, different process forces occur that could lead to varying tool deflections and defects of the cutting edge. Knowledge about the relation between grinding strategy and material removal rate can lead to improved work results in the form of higher preciseness of shape and dimensional accuracy as well as enhanced cutting edge quality. CAD/CAM-programs do not provide a proper calculation of chip cross-sectional areas and thus material removal rates. Hence, it is necessary to develop suitable simulation models.

To enable a reproducible manufacturing of cutting edge geometries, different strategies for form grinding of cutting edge microgeometries, with regard to material removal rates by using permeation simulations, are examined. In grinding tests, the spindle power for the different grinding strategies is compared with the course of the graphs of the material removal rate. The aim of this analysis is to assess these different form grinding strategies with regard to material removal rates to increase the process performance and work results.

2. Test Conditions

The program to simulate the permeation of the tool and the workpiece was developed with the program software MATLAB, The MathWorks Inc., Natick, MA, USA. The form grinding processes of the cutting edge micro geometries were executed on a five-axis-machining center RXP600DSH of Röders GmbH, Soltau, Germany. Besides three linear axes (X, Y, Z), this machine is equipped with two rotatory axes (A, C). Furthermore, the integrated grinding spindle has a maximum power of 8.5 kW to enable a maximum rotation speed of 60,000 rpm. To dress the grinding tools, a dressing spindle with a sintered diamond form roller is integrated. To produce the required sharpness of the mounted points, a sharpening block is used and an automated sharpening process using the through-feed method was performed. Moreover, the grinding machine enables the determination of the geometry and the reference of the part with the help of a 3D-probe. The workpieces were cutting inserts made of cemented carbide of the type SNMA120408. As cooling lubricant, grinding oil of MKU-Chemie GmbH, Rödermark, Germany, with a viscosity of 7.6 mm^2/s at 40 °C was used. Moreover, metal bonded diamond mounted points with a grain size of D20 were utilized. Only the use of diamond tools allows an economic grinding processing of cemented carbides. To reduce tool wear and thus dimensional deviations on the workpiece, metal bond tools were used. An important advantage of metal bonds is the shape stability of the tool profiles [9]. In order to gain process forces, conventional force measurement platforms cannot be satisfactorily used when grinding with 5-axis kinematics because of relocations of the self weight. Rotating multicomponent dynamometers could be suitable solutions but the non-contact signal transmission caused by the design principle can be used only at maximum spindle speeds of approximately 12,000 rpm. Therefore, the performance data of the grinding spindle was recorded and used to validate the simulation results. To verify the accuracy of this measurement method, the performance data was compared to force values recorded by a force measurement platform during a peripheral longitudinal grinding process, as shown in Figure 1.

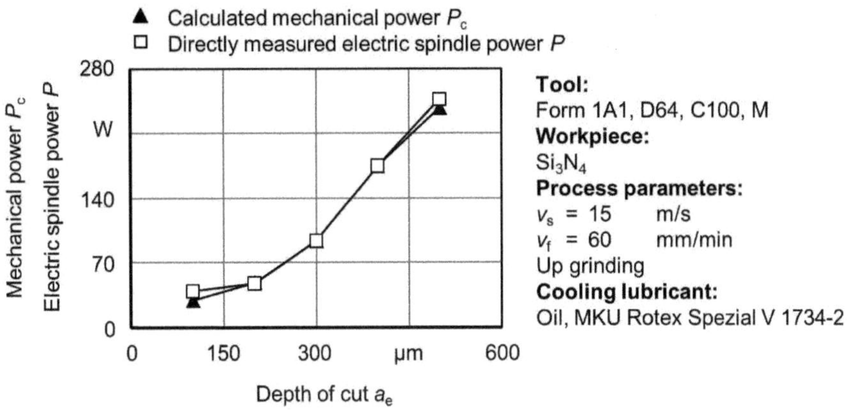

Figure 1. Comparison of calculated mechanical power P_C with directly measured electric spindle power P.

It is shown that, with the used process parameters, both force progressions, from depths of cut of approximately 200 μm, show a good correlation. During manufacturing, cutting edge microgeometries with much smaller material removal rates than in Figure 1 will be realized. To create equally reliable statements between spindle power data and process forces, the material removal rate was increased by offsetting the target geometry in the workpiece, Figure 2.

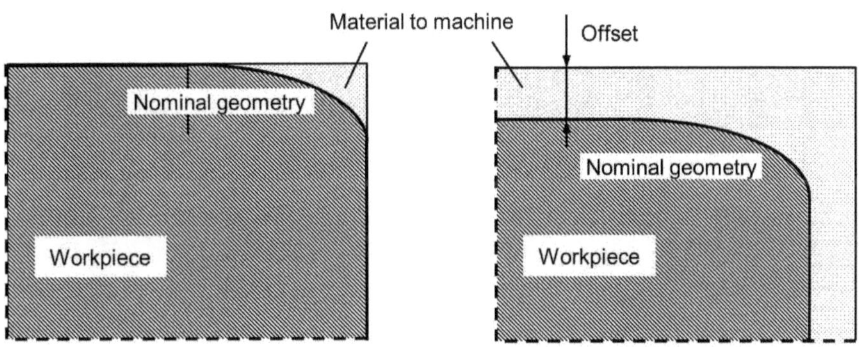

Figure 2. Offset of nominal geometry to increase material removal rates.

Due to the relatively long tool paths of the mounted point with a constant depth of cut on the rake and flank face, these constant values of spindle power can be seen as the reference during grinding of the cutting edge. Therefore, it is possible to identify the actual machining of the cutting edge. The difference between the spindle power minimum and maximum is shown in the further graphs as a percentage increase of spindle performance in comparison to the minimum.

3. Kinematic-Geometric Model

3.1. *Cutting Edge Microgeometries and Tool Types*

In the investigations, two different cutting edge geometries were considered: an ideal radius and a waterfall radius. Both microgeometries can be characterized using the form-factor method. The form factor is the relationship between the cutting edge segment on the rake face S_γ and the cutting edge segment on the flank face S_α. The ideal radius has a form factor $\kappa = 1$. Therefore, the cutting edge segments on flank and rake faces, as well as the cutting edge radius, have the same value. The waterfall radius is characterized by a form factor greater than one, in this case $\kappa = 1.5$, Figure 3.

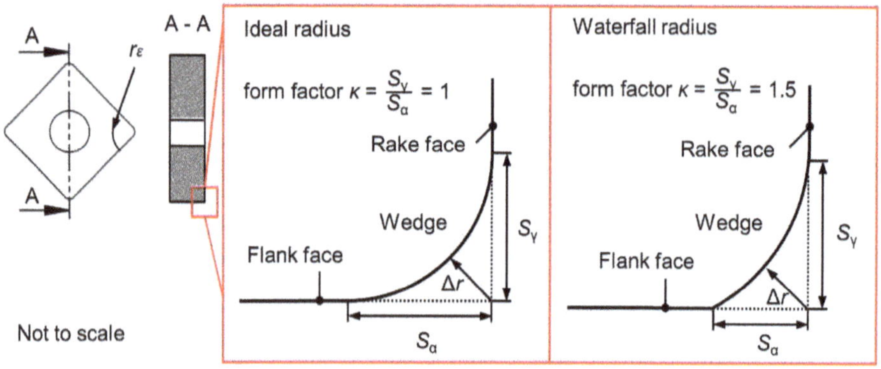

Figure 3. Form-factor method for ideal radius and waterfall radius.

For both cutting edge microgeometries, different grinding strategies were examined, as shown in Figure 4. Grinding strategy I and II differ with regard to the active tool shape. In Strategy I, the sphere-shaped form of the cylinder ball end mounted point is used to cut orthogonally above the cutting edge, referred to as the course of the corner radius r_ε. Strategy II has the same tool kinematics but the cylindrical part of the mounted point is utilized for the cutting process. In Strategy III, the ball-shaped part of the mounted point is guided above the cutting edge at an angle of 45°, referred to as the course of the corner radius. For machining the waterfall geometries with a form factor $\kappa = 1.5$, the grinding strategies I and II match with the strategies of the ideal grinding radius.

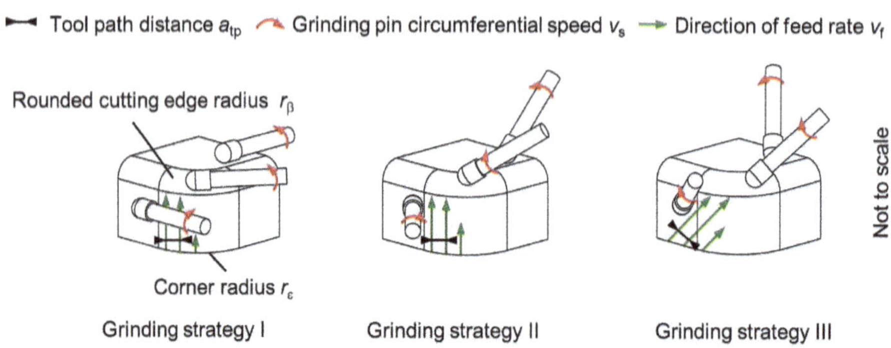

Figure 4. Different grinding strategies for NC-form grinding cutting edge microgeometries.

3.2. *Modelling of the Permeation*

The aim of the developed process simulation is to identify the material removal rates of different grinding strategies and cutting edge microgeometries for one tool path movement. Thus, knowledge about forces involved and therefore about tool deflection can be generated to analyze different grinding strategies. The kinematic-geometrical model simulates the penetration of two geometrical bodies. The workpiece (cutting insert) is defined as a cylindrical surface and the mounted point is also defined as a cylindrical surface if the required tool type is cylindrical; if the required tool type is spherical, an appropriate partial surface is created, Figure 5. The geometric objects are described as point clouds.

To reduce the computing time, the program minimizes the point clouds to the points of the contacting areas of both geometric objects.

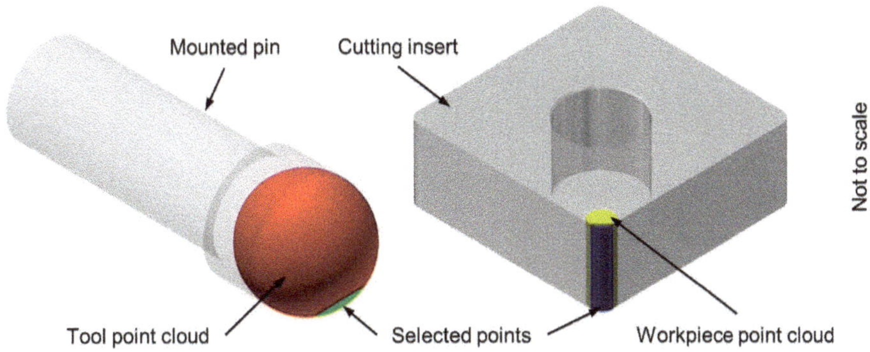

Figure 5. Schematic representation of the tool and workpiece areas relevant for simulated calculation.

The target geometry of the respective cutting edge microgeometry is precisely positioned as a numerical approximated curve in the workpiece point cloud. This curve is the reference for the radial position of the most external point of the tool point cloud. For each calculation step, the tool point cloud is placed at the following point of the cutting edge curve, from the flank face (angle position $\alpha = 0$) to the rake face (angle position $\alpha = 90°$). Therefore, the rotation of the approximated tool around the approximated cutting edge enables the determination of the exact section planes. In Figure 6, the green cycles represent the reduced tool point cloud and the blue area represents the reduced workpiece point cloud.

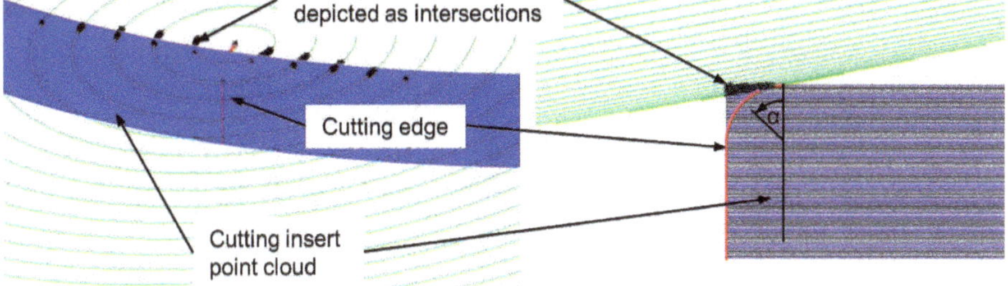

Figure 6. Intersection of tool and workpiece point clouds in one position on one tool path.

The red curve depicts the cutting curve of the cutting edge target geometry. The point distance of both point clouds is 0.1 μm to achieve high calculation accuracy. For each position of the tool point cloud, the intersection points are projected on a plane. Afterwards, the area within the external points can be determined which is equivalent to the section plane between the tool and workpiece, Figure 7.

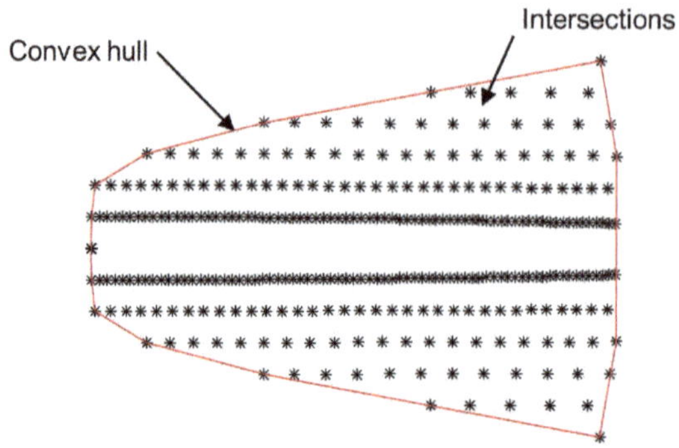

Figure 7. Generating the cut surface by locating the outer intersections.

Before running the simulation, the boundary conditions have to be defined. These include geometry data of the cutting edge microgeometry and the tool. Further simulation data is the feed speed and the resolution of geometric objects influencing the accuracy of the calculations.

4. Results

4.1. Simulation Results

The results of the simulation show the different material removal rate Q_w curve progressions for each grinding strategy, as shown in Figure 8. The lowest maximum material removal rate $Q_{w,max}$ is achieved by grinding strategy II.

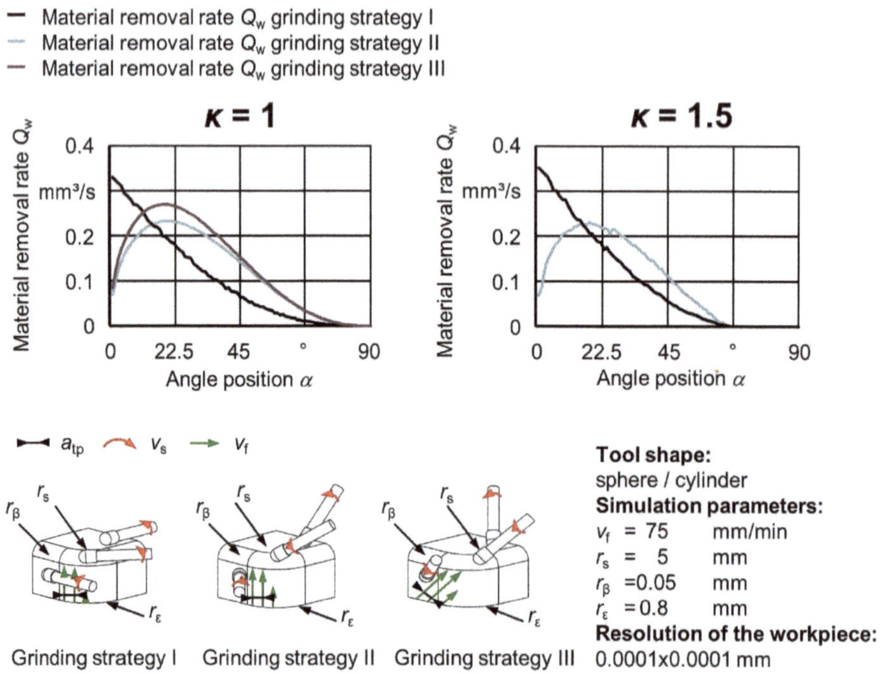

Figure 8. Simulation results for different form grinding strategies.

The maximum value is located at an angle position α of approximately 20°. Therefore, the maximum material removal rate $Q_{w,max}$ of grinding strategy I was determined at the start angle position of the mounted point, referred to as the course of the rounded cutting edge radius r_β. It is the highest achieved value. Following the further angle positions, the curve is regressive and has low values from angle position $\alpha = 16°$. The course of grinding strategy III is similar but has a higher peak than grinding strategy II. Therefore, it does not appear advantageous for process forces and consequently for tool deflections. Hence, further considerations relate to grinding strategies I and II. The courses of the different grinding strategies I and II for grinding waterfall radii with a form factor $\kappa = 1.5$ differ only slightly from the courses of the equivalent grinding strategies for the ideal radii, Figure 8. With an angle position of $\alpha = 70°$, almost all the material is removed. This is caused by the larger curvature at the beginning of the cutting edge geometry and by the fact that the ratio of the tool radius to the cutting edge radius is approximately 100. This results in relatively flat tool geometry, as shown in Figure 6.

4.2. Experimental Results

To validate the kinematic-geometrical model, grinding tests were conducted. For this purpose, the spindle power of the different grinding strategies was recorded and compared with the courses of the material removal rates from the simulation calculations, as shown in Figure 9. Generally, for all test series, the courses of the spindle power are highly similar to the courses of the simulation calculations. However, small course deviations can be identified.

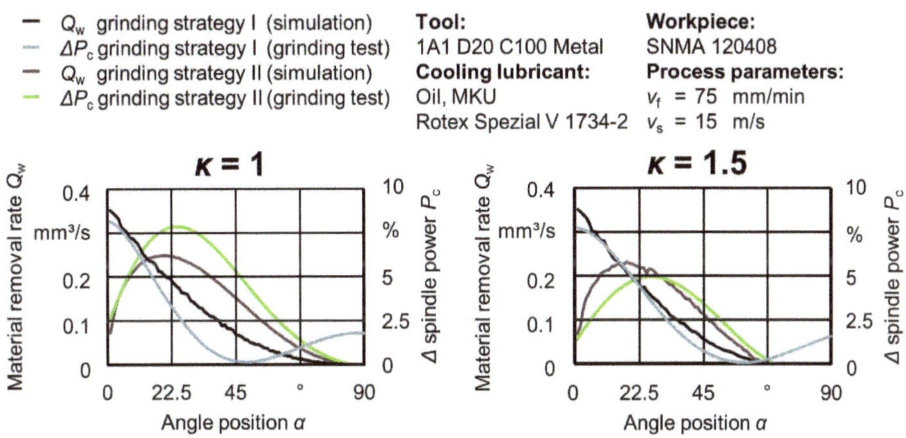

Figure 9. Comparison of curve progressions of simulation data and spindle power data.

For both cutting edge geometries, slight deviations in the spindle power courses can be observed in comparison to the simulation results, Figure 9. The reason for this occurrence may be the inertia of the tool spindle's continuous turn regulation in combination with the comparatively short process time of one tool path of approximately 8 s. Another important factor is the relatively small spindle load and therefore the less linear behavior between the material removal rate and process forces, as well as the correct determination of the relevant part of the spindle power course. Nevertheless, analysis of the spindle power data obtained in the grinding tests confirms the trend of the simulation results. The validation of the simulation model confirms the benefit of form grinding processes of cutting edge microgeometries for planning strategies. The kinematic-geometrical model enables the prediction of maximum material removal rates and subsequently indicates process forces and potential tool deflections. Furthermore, the influence of the grinding strategy on possible damages to the cutting edge can be analyzed.

5. Summary

Within the presented work, a kinematic-geometric model was developed in order to analyze the complex contact conditions during grinding cutting edge microgeometries. Therefore, a simulation approach based on the intersection of geometric bodies was used. The grinding tool and workpiece surface were defined as cylindrical point clouds and the cross section of the cutting edge microgeometry was determined by numerical approximation. By moving the tool point cloud stepwise along the coordinates of the cutting edge microgeometry for each position, the interface between the workpiece point cloud was calculated. Thus, the varying material removal rates could be determined and compared with the spindle power data. The relatively even curve progressions show the correlation between the material removal rate Q_W and the process forces. Therefore, the geometric-kinematic model can be used to analyze different form grinding strategies for manufacturing various cutting edge microgeometries with regard to process forces and consequently tool deflection. With the gathered information, the work results of form grinding cutting edge microgeometries, regarding the form accuracy and the chipping of the cutting edge, can be improved.

Acknowledgments: The authors would like to thank the Deutsche Forschungsgemeinschaft (German Research Foundation) for funding this research within the project DFG UH 100/179-1.

Author Contributions: Both authors were equally involved in creating this paper.

Conflicts of Interest: The authors declare no conflict of interest.

References

1. Biermann, D.; Wolf, M.; Aßmuth, R. Cutting edge preparation to enhance the performance of single lip deep hole drills. *Procedia CIRP* **2012**, *1*, 172–177. [CrossRef]
2. Klocke, F.; Kratz, H. Advanced tool edge geometry for high precision hard turning. *CIRP Ann. Manuf. Technol.* **2005**, *54*, 47–50. [CrossRef]
3. Denkena, B.; Biermann, D. Cutting edge geometries. *CIRP Ann. Manuf. Technol.* **2014**, *63*, 631–653. [CrossRef]
4. Kümmel, J.; Braun, D.; Gibmeier, J.; Schneider, J.; Greiner, C.; Schulze, V.; Wanner, A. Study on micro texturing of uncoated cemented carbide cutting tools for wear improvement and built-up edge stabilization. *J. Mater. Process. Technol.* **2015**, *215*, 62–70. [CrossRef]
5. Denkena, B.; Grove, T.; Bergmann, B. Eine frage des radius. *WB* **2016**, *149*, 58–61.
6. Uhlmann, E.; Oberschmidt, D.; Löwenstein, A.; Polte, M.; Winker, I. Schneidkantenpräparation von VHM-Mikrofräsern. wt Werkstattstechnik Online 2015. Available online: http://www.werkstattstechnik.de/wt/article.php?data%5Barticle_id%5D=84726 (accessed on 30 May 2017).
7. Gleim, P. *Untersuchungen Zum Bandsägen Mit Diamantbeschichteten Werkzeugen*; Kassel University Press GmbH: Kassel, Germany, 2006.
8. Schedler, W. *Hartmetall Für Den Praktiker: Aufbau, Herstellung, Eigenschaften Und Industrielle Anwendungen Einer Modernen Werkstoffgruppe*; VDI-Verlag: Düsseldorf, Germany, 1988; ISBN 3184008037.
9. Denkena, B.; Tönshoff, H.K. *Spanen Grundlagen*; Springer: Heidelberg, Germany; Dordrecht, The Netherlands; London, UK; New York, NY, USA, 2011; ISBN 978-364219772-7.

inventions

MDPI

Article

Influence of the Grinding Wheel Topography on the Thermo-Mechanical Stress Collective in Grinding

Sebastian Barth * and Fritz Klocke

Laboratory for Machine Tools and Production Engineering (WZL), RWTH Aachen University, 52074 Aachen, Germany; f.klocke@wzl.rwth-aachen.de

* Correspondence: s.barth@wzl.rwth-aachen.de; Tel.: +49-241-8028183

Received: 16 October 2017; Accepted: 29 November 2017; Published: 4 December 2017

Abstract: The grinding process is used for both high-performance machining and surface finishing of hardened steel. In addition to the grinding parameters and the grinding fluid supply, the topography of the grinding wheel mainly determines the grinding process behavior and the grinding process result. An alteration of the topography by a variation of the volumetric composition of the grinding wheel, by a variation of the grinding wheel conditioning, or by wear causes a change in the contact conditions. The state of the art shows a substantial knowledge deficit about the influence of the volumetric grinding wheel composition and the resulting grinding wheel topography on the thermo-mechanical stress collective acting on the workpiece external zone. Thus, it is not possible to make a quantitative statement about the influence of the volumetric grinding wheel composition on the external zone properties of a component after grinding. Therefore, the aim of the current research is an empirical-analytical model for the prediction of the thermo-mechanical stress collective as a function of the grinding wheel topography. For this purpose, a methodology is developed, which enables the prediction of the topography-dependent thermo-mechanical load in a grinding process. Therefore, the topography is characterized by means of quantitative parameters and the main influencing variables on the grinding process behavior are investigated. The findings are used to analyze the influence of a change in the topography on the grinding temperature and the grinding force. The obtained results are summarized and are used to explain the thermo-mechanical stress collective as a function of the grinding wheel topography.

Keywords: grinding; topography; thermo-mechanical stress collective; cutting edge; contact conditions

1. Introduction

The topography of a grinding wheel represents a functional surface. It is part of a tribological system and causes a stress collective during the engagement with the workpiece material. It determines the chipping of the workpiece material, and thus, the stresses, such as mechanical loads and temperatures. To identify the influence of the grinding wheel topography on this thermo-mechanical stress collective, the topography of the grinding wheel must be described adequately. In principle, the grinding wheel topography is dependent on the grinding wheel specification [1,2]. Further, it is strongly influenced by the grinding-wheel conditioning process and geometrical changes during the grinding process due to wear [3,4]. The characteristics of the topography determine the machining behavior in the grinding process, particularly, so it must be taken into account when designing grinding processes [5].

In the past, several approaches were developed to characterize the grinding wheel topography and to investigate its influence on the grinding behavior. Nowicki proposed a method for selecting topographic characterization parameters for a multi-parameter representation based on low statistical correlation analysis. Based on 100 machined and ground surfaces, he constructed one cross-correlation matrix. Nine parameters were compared, including three height parameters, four spacing parameters, asperity radius, and slope. The three height parameter pairs were found to have correlation coefficients above 0.9, but for the other 29 pairs, he did not find any correlation [6]. Hecker et al. presented a three-dimensional methodology to evaluate the static parameters of a topography to calculate the dynamic parameters. In their model, the individual grain geometries were approximated to a conical or a spherical shape. They found that primarily the protrusion of the grains determines the dynamic grain density and the grinding force [7]. For the characterization of the topography, Nguyen and Butler used the root-mean square parameter Sq to describe its coarseness and the parameters density of summits Sds and summits curvature Ssc to indicate the abrasive grain density and their sharpness. The results show a significant correlation between the parameters Sds and Sq, but the results do not show a correlation between the topography and the investigated grinding force [8]. Uhlmann et al. described the surface of a grinding wheel to develop a reliable method for Numerical Control (NC)-grinding by using a single profile of kinematic cutting edges instead of the whole topography. The simplification of the grinding tool topography proved to be feasible for grinding with abrasive mounted points, in which rotations are high and feed rates are low. A quantitative description of the topography or the kinematic cutting edges is not given [5].

According to Hübert et al., the sufficient characterization of the grinding wheel topography is only possible with the aid of value-supported height, surface, and volume data [9]. Since 2007, DIN ISO 25178 standard [10] has consistently been used to control the acquisition and determination of surface-related and three-dimensional surface parameters. In order to characterize a surface, it is differed between surface-related height parameters and functions, including related function parameters.

The height parameters include the mean arithmetic height of a surface Sa, which indicates the arithmetic mean value of the absolute ordinate values within a definition range, and the mean square height of a surface Sq, which calculates the root from the average square value of the ordinal values of a range. The skewness of a surface Ssk, which corresponds to the quotient of the mean third power of the ordinate values and the third power of Sq, is a measure of the symmetry of the amplitude density curve. Furthermore, the kurtosis of the surface Sku, the steepness of the height distribution within the definition area, as well as the maximum peak and maximum sink height Sp and Sv, and the maximum height Sz (sum of Sp and Sv), are counted to the height characteristics.

Furthermore, the standard specifies the definition of functions for describing the surface texture and the related parameters. In particular, the area-based material ratio curve of scale-limited surfaces, also known as the Abbott-Curve, is of interest in the description of the grinding wheel topography. This function represents the material portion of the scale-bound surface as a function of height, and thus describes the increase in the proportion of material (Mr) of a surface with the increasing depth of the roughness profile.

The Abbott-Curve is formed by cutting the profile of a grinding wheel from the highest peak to the deepest valley at defined intervals from a horizontal plane. Important components of the Abbott-Curve are the core height Sk, which corresponds to the distance between the highest and lowest levels of the core of the surface, the reduced peak height Spk, which indicates the mean height of the protruding peaks above the core height, and the reduced valley height Svk, which indicates the prominent valleys below the core.

The Abbott-Curve also opens the possibility to calculate the area-related empty volume Vv(p) of the cavities for a given material proportion p. Based on this, the empty volume of the valleys of a surface Vvv and the empty volume of the core Vvc is calculable. Analogously to the calculation of the empty volume, the peak material volume Vmp is calculable for a given material proportion. However, the Abbott-Curve does not provide information about the ratio of the proportions of bond and abrasive grains on the surface of a grinding wheel [11].

Weiß, Duscha, and Rasim pursued in their research a different approach for the description of the grinding wheel topography. They quantitatively described the contact conditions in grinding by means of a grain engagement model, which bases on the limiting cutting edge offset angle ε_{lim} according to Kassen and Werner [12,13]. Therefore, they developed the software Topo-Tool, which focuses on the analysis and quantitative description of the topography by analyzing the kinematic cutting edges. Weiß [14] determined the number of theoretical kinematic grain surfaces N_{kin}, as well as their respective size, A_{kin}, and calculated the fractions of these surfaces in the tangential and normal direction to the grinding direction $A_{kin,t}$ and $A_{kin,n}$. He quantified the engagement surfaces of measured topographies of ceramic and synthetic resin bonded grinding wheels depending on their specification, as well as the average tangential and normal cutting engagement surfaces $\bar{A}_{kin,t}$ and $\bar{A}_{kin,n}$ as a function of the mean grain diameter [15]. Duscha used the kinematic engagement surfaces of a grinding wheel to determine the maximum surface pressure in the grinding process. In this way, he was able to draw conclusions about the real mechanical loads in the grinding process [16]. Rasim enhanced the Topo-Tool model with the four additional characteristic values peak angle β, rake angle γ, and sharpening angle δ, as well as the opening angle α to describe the kinematic cutting engagement surfaces. He determined the frequency distribution of the grain shapes, which he used to analyze the influence of the shape of the cutting edge surfaces on the energy conversion in the grinding process. In his investigations, he varied the grain type in galvanic bonded grinding wheels, and was thus able to model the influence of the grinding wheel topography on the energy conversion in the grinding process for the first time [17]. However, he did not undertake a comprehensive variation of the volumetric grinding wheel composition. Figure 1 gives an overview of the potential parameters for describing the properties of functional surfaces and grinding wheel topographies.

Despite the previous research, the state of the art shows a substantial knowledge deficit about the influence of the volumetric grinding wheel composition, and the resulting grinding wheel topography on the thermo-mechanical stress collective acting on the workpiece external zone. Thus, it is not possible to make a quantitative statement about the influence of the topography geometry on the thermal and mechanical process load acting on the workpiece external zone of a component. Consequently, the knowledge about the influence of the topography of a grinding wheel on the component's surface and external zone properties is insufficient.

Therefore, the aim of current research is an empirical-analytical model for the prediction of the thermo-mechanical stress collective as a function of the grinding wheel topography. For this purpose, a methodology has been developed that enables the prediction of the topography-dependent thermo-mechanical load in the grinding process.

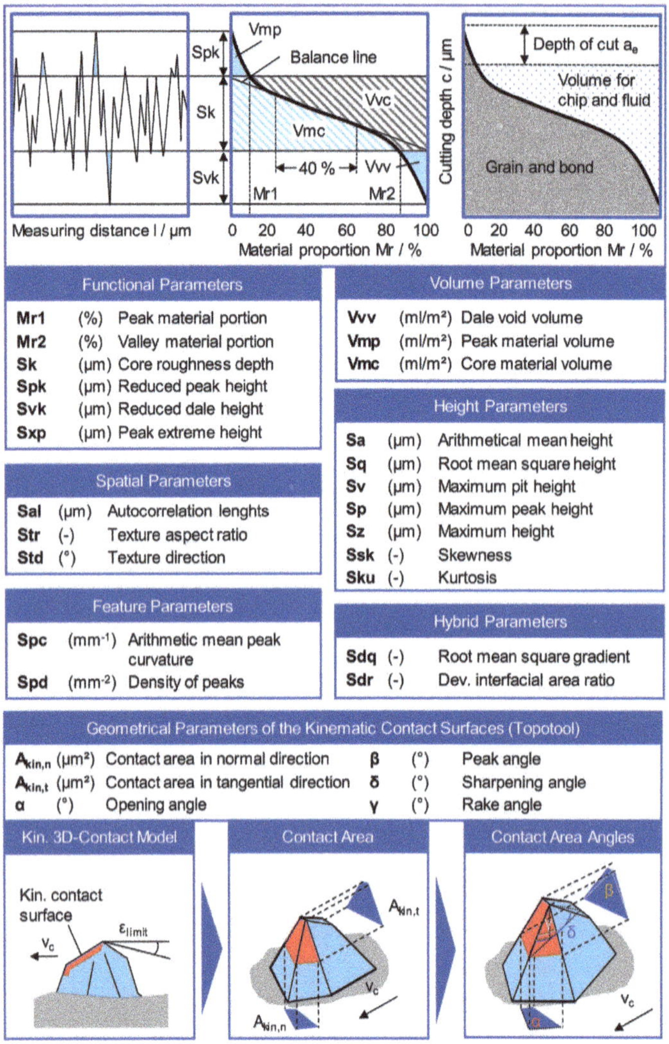

Figure 1. Parameters for the description of the properties of functional surfaces and grinding wheel topographies.

2. Materials and Methods

First, grinding tests with multilayered, synthetic resin-bonded cubic boron nitride (CBN) grinding wheels have been carried out. In the grinding experiments, a tool grinding machine S22P turbo of the company ISOG was used to investigate the thermo-mechanical stress collective depending on the grinding wheel topography. During the experimental investigations, 13 synthetic resin-bonded CBN grinding wheels from the company TYROLIT Schleifmittelwerke Swarovski K.G. (Schwaz, Austria) were used to grind the bearing steel 100Cr6. The grinding wheels differed in their volumetric composition, in the average grain size, and in the grain type. The grain types ABN200 and ABN800 from the company Element Six (Burghaun, Germany) were used. The grain type ABN200 is described as "sharp" with high strength and thermal stability. The grains predominantly have an octahedral shape and are rounder than the grains of the grain type ABN800. An overview of the used grinding wheel specifications is shown in Figure 2.

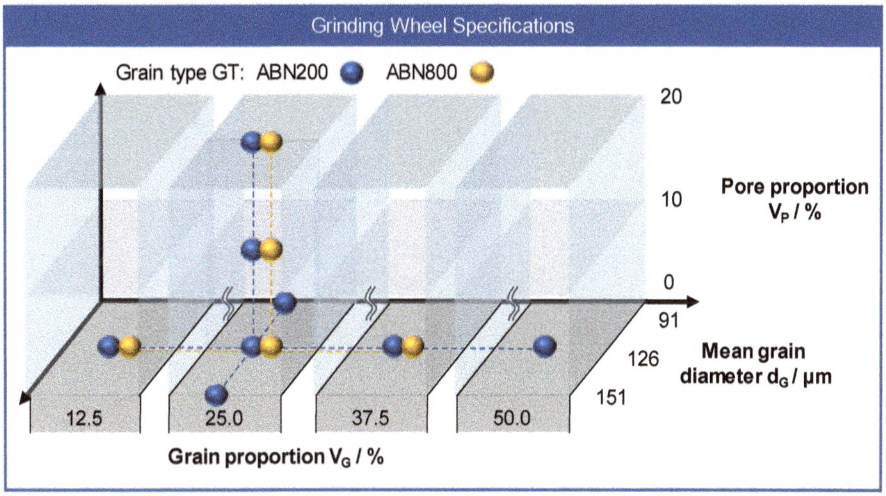

Figure 2. Overview of the used grinding wheel specifications.

Approximately 160 grinding tests were carried out, each with different experimental conditions to investigate the influence of the grinding wheel topography on the thermo-mechanical stress collective. A full-factor experimental design was chosen in order to fully explore the influence of the individual input factors. By varying only one input variable at otherwise constant input variables, it was possible to identify the influence of each input variable on the process result. Thus, almost-repeated tests were carried out in the experimental design, which could be confirmed on the basis of the results. Both, the composition of the grinding wheels and the dressing process were varied to adjust the topography of the grinding wheels. The experiments were carried out in an up-grinding process. In the experiments, the grinding wheel topography was varied by changing the volumetric composition of the grinding wheels. The CBN abrasive wheels were profiled by a diamond form roller with a mean grain size of 301 µm and were sharpened by means of sharpening blocks of the type A240 H5 V from TYROLIT Schleifmittelwerke Swarovski K.G. company. They were dressed with different dressing overlap ratios $U_d = 1; 2; 4$, a constant dressing feed $a_{ed} = 2.5$ µm and a speed ratio of $q_d = 0.8$. In addition, the topography was varied by using the sharpening parameters specific sharpening material removal rate $Q'_{Sb} = 100$ mm^3/mms and specific sharpening material removal $V'_{Sb} = 50$, 100 and 150 mm^3/mm. The grinding parameters grinding wheel circumferential speed $v_s = 30$ m/s and infeed $a_e = 100$ µm were kept constant. The workpiece speed was varied in three steps between 1.5 m/min < v_w < 4.5 m/min to set the specific material removal rate varied between 2.5 mm^3/mms < Q'_w < 7.5 mm^3/mms.

The interpretation of the thermo-mechanical stress during grinding is only possible with a sufficiently precise knowledge of the occurring grinding force and temperature, as well as the geometric contact conditions along the real contact arc. Therefore, a temperature-force measuring platform has been manufactured to measure the occurring grinding forces and temperatures. In order to determine the thermal loads during the grinding process parallel to the mechanical loads, a temperature measurement method according to Duscha, based on the work of Choi and Batako, was used [16,18,19]. A constantan foil (55% Cu, 44% Ni and 1% Mn) was connected to a constantan wire with the help of a TL-WELD-220V thermocouple welding machine from OMEGA. The resulting metallic conductor was fixed between two screwed workpiece parts. Ceramic mica discs isolated the constantan foil from the two workpiece parts beside. The downstream workpiece part was connected with an iron conductor to a reference point. As a result of the passage of the grinding wheel during the grinding process, the constantan foil was machined, and, due to plastic deformation, a contact with the downstream part of the workpiece was closed over the isolation. The closed contact between the metal pair enabled the temperature measurement due to the occurring thermal stress by means of the electrical charge transfers in a high repeatability and certainty of results.

To ensure that the initial wear of the grinding wheels was overcome before the grinding experiments, each grinding wheel was ground in after the dressing process. Once the grinding wheel produced a constant grinding force, the grinding in process was stopped.

In order to identify the influence of the grinding wheel topography on the thermo-mechanical stress collective, the topography of the grinding wheels was recorded and analyzed by a three-dimensional (3D) Laser Scanning Confocal Microscope VK-X 150 before each experiment. The analysis of the topographies was carried out with the analysis module of the company Keyence (Osaka, Japan) and with the MATLAB-based Topo-Tool developed at the Laboratory for Machine Tools *and* Engineering (WZL) of RWTH Aachen University [14]. The analysis module of the laser scanning microscope enables the determination of the volume, height, distance and function parameters listed in Figure 1 according to DIN EN ISO 25178 [10]. The Topo-Tool enabled the identification of the kinematic engagement surfaces of a grinding topography, as well as the calculation of the size of the engagement surfaces, depending on the process parameters. The calculation is based on the Formula 1 calculated by Kassen and Werner for the calculation of the limiting cutting edge angle ε_{lim} [12]. The calculated limiting cutting edge angle ε_{lim} results from the infeed of the grinding wheel during simultaneous relative movement between the grinding wheel and the workpiece in the circumferential direction of the grinding wheel. In Figure 3, the approach for calculating the kinematic engagement surfaces is shown schematically, using the example of a small grinding wheel cutout.

$$\varepsilon_{lim} = \frac{v_w}{v_s} \cdot 2 \cdot \left(\frac{a_e}{d_s}\right)^{\frac{1}{2}} \tag{1}$$

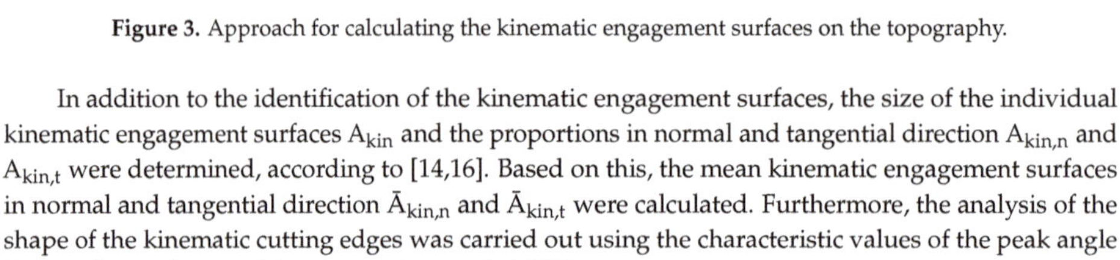

Figure 3. Approach for calculating the kinematic engagement surfaces on the topography.

In addition to the identification of the kinematic engagement surfaces, the size of the individual kinematic engagement surfaces A_{kin} and the proportions in normal and tangential direction $A_{kin,n}$ and $A_{kin,t}$ were determined, according to [14,16]. Based on this, the mean kinematic engagement surfaces in normal and tangential direction $\bar{A}_{kin,n}$ and $\bar{A}_{kin,t}$ were calculated. Furthermore, the analysis of the shape of the kinematic cutting edges was carried out using the characteristic values of the peak angle β, the rake angle γ, and the sharpening angle δ [17].

3. Results

The aim of the research was to investigate the influence of the grinding wheel topography on the thermo-mechanical stress collective. Therefore, first the influence of the listed topography parameters on the grinding force and the grinding temperature was analyzed. Parameters with an influence on the thermo-mechanical stress collective were subsequently tested for correlations with each other and were then grouped. Subsequently, the influence of a change of the relevant topography parameters on the process results was examined in detail. The high number of experiments and investigated parameters did not permit any direct analysis and interpretation of all the results. Therefore, first investigations that are presented in this paper were carried out with a constant specific material removal rate of $Q'_w = 2.5\ mm^3/mms$, a constant dressing overlap ratio $U_d = 1$ and with grinding wheels with a grain proportion of $V_K = 25\%$ and mean grain size of $d_G = 126\ \mu m$.

The analysis of the test results was carried out with the aid of the open-source software ORANGE, which is a component-based visual programming software package for data analysis, data visualization, machine learning, and data mining [20]. The visualization of the correlations between topography parameters and process parameters was carried out using scatter plot matrices in MATLAB. This enabled the rapid identification of correlations between the parameters.

3.1. Correlations between the Process Result Parameters

Figure 4 shows the correlation between the maximum temperature T_{max} and the grinding force in tangential and normal direction F_t and F_n for the specific material removal rate $Q'_w = 2.5\ mm^3/mms$. In the present case, the highest value for the degree of determination R^2 was found with a linear approach to the relationship between the grinding force and the temperature. In particular, there was a very strong linear correlation between the maximum temperature T_{max} and the grinding forces in normal and tangential direction F_n and F_t. Based on this finding, the influence of the topography on the process behavior was investigated by means of the grinding temperature.

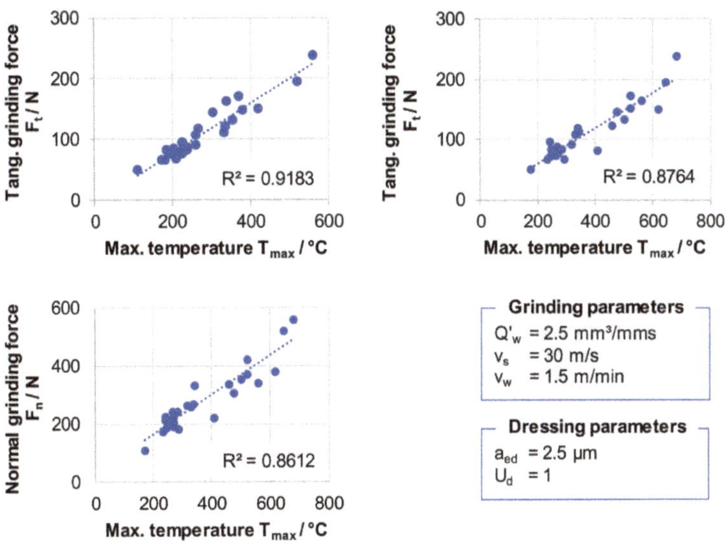

Figure 4. Correlation between the process force and the process temperature.

3.2. Correlations between the Topography Parameters

In the following, the characteristic topography parameters, which had a similar influence on the thermo-mechanical stress collective were investigated and grouped. Three parameter groups were identified by the analysis, so that a reduction of the total of 33 investigated topography parameters to three parameter groups could be realized for describing the influence of the grinding wheel topography on the thermo-mechanical stress collective.

The results did show no correlation between the height parameters Sa, Sz, Sq, Ssk, Sku, Sp, and Sv among each other and the process results maximum temperature T_{max} and the process force components F_n and F_t. The process results also showed no significant dependencies on the height parameters Str, Sal, and Std. Therefore, height parameters are not suitable for describing the influence of the grinding wheel topography on the thermo-mechanical stress collective.

The volume parameter texture aspect ratio of the surface Str, autocorrelation length of the texture Sal, and the texture direction Std, describe the spatial structure patterns of the topography. Significant dependencies of the process variables were not found in the investigations, so that the volume parameters do not seem to be suitable for the quantitative description of the influence of the topography on the grinding process behavior.

In mechanical engineering, hybrid parameters are used in various applications to describe wear and tribological behavior. They combine spatial and vertical parameters. The developed interfacial area ratio Sdr indicates the ratio between the true surface of the topography and the projected base area in the *X*–*Z* plane. Thus, it is a measure of the three-dimensional roughness of the analyzed surface. A perfectly planar surface has a value of Sdq = 0. The arithmetic mean value of the peak curvature Spc, the developed interfacial area ratio Sdr, and the root mean square gradient Sdq showed a particularly strong correlation. The correlation between the arithmetical mean value of the peak curvature Spc and the developed interface area ratio Sdr was best describable by a quadratic function with a coefficient of determination of R^2 = 0.9589, which corresponds to an excellent correlation (see Figure 5). The correlations between the above-mentioned parameters and the process variables were investigated and they showed identical trends. For the further procedure, the three parameters were grouped and the influence of the arithmetical mean value of the peak curvature Spc was analyzed as representative for the parameter cluster.

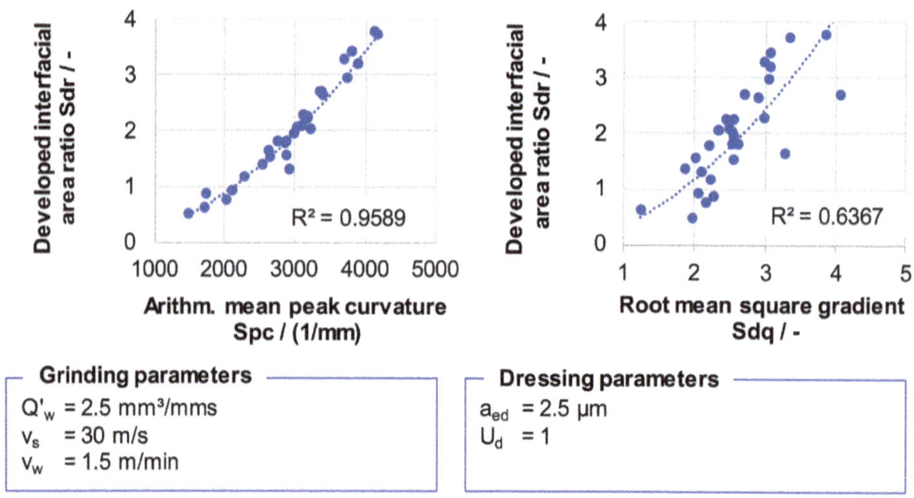

Figure 5. Correlation between the developed interfacial area ratio, the arithmetic mean peak curvature and the root mean square gradient.

Similarly, there were strong correlations between the functional parameters describing the upper part of the Abbott curve, see Figure 6. Since the correlation coefficients between the reduced peak height Spk, the peak extrem heigt Sxp, and the peak material volume Vmp assume approximately a coefficient of determination of nearly R^2 = 1, an almost perfect linear relationship exists between the topography characteristics. Here, the formation of a cluster is obvious, since all the parameters also had the same relation to the process variables.

The angles of the kinematic cutting edges, which were identified by the Topo-Tool, also correlated strongly. Figure 7 shows approximately linear relationships between the angles of the kinematic cutting edge areas. When the mean peak angle $\overline{\beta}$ or the mean sharpening angle $\overline{\delta}$ increase, the mean rake angle $\overline{\gamma}$ decreases. Here, too, a grouping of these parameters was appropriate, in order to reduce complexity for the analysis of the influence of the grinding wheel topography on the thermo-mechanical stress collective.

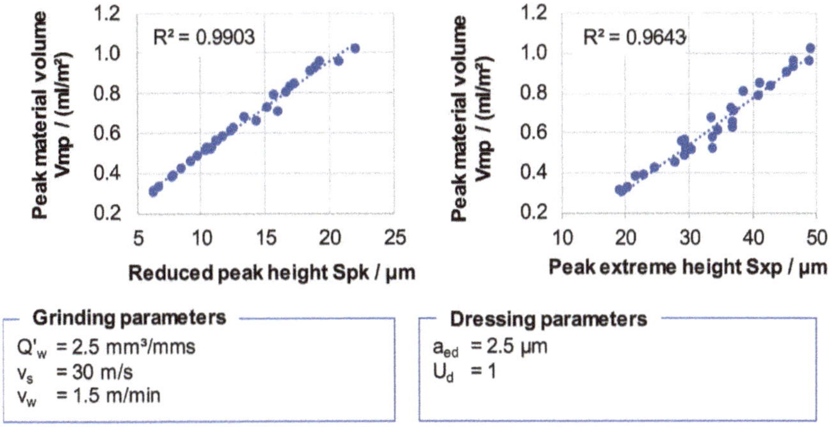

Figure 6. Correlation between the peak material volume, the reduced peak height and the peak extreme height.

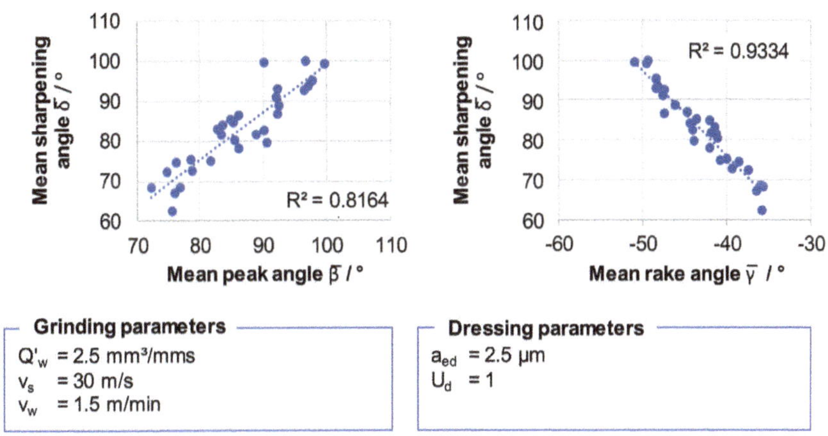

Figure 7. Correlations between the angles of the kinematic cutting edge areas.

3.3. *Influence of the Topography Parameters on the Chip Thickness and Process Result Parameters*

Based on the rapid identification of correlations between the topography parameters and the process parameters, a detailed analysis of the influence of the relevant topography parameters on the thermo-mechanical stress collective was carried out. Figure 8 shows the correlations between the hybrid parameters and the process parameter mean maximum chip thickness $\overline{h}_{cu,\,max}$. The maximum chip thickness was investigated by means of the Topo-Tool, which measures the biggest vertical distance between the lowest and the highest points of the tangential kinematic engagement area, as described in Figure. Between the developed interface ratio Sdr and the process parameters F_n, F_t, and T_{max}, only weak correlations were identified. An influence of these surface parameters on the process behavior is physically close, but due to the coefficient of determination of only $R^2 = 0.62$, a low suitability of these topography characteristics is assumed to describe the influence of the topography on the thermo-mechanical stress collective solely.

The further investigated functional parameters are derived from the Abbott curve, which characterizes the vertical material distribution of a surface structure. The Abbott curve is created by applying the material portion above the profile height. Functional parameters, such as the reduced dale height Svk, or the core material volume Vmc, which describe the geometry and structure of the grooves on a surface, did not influence the force or temperature in the grinding process. In particular, the reduced dale height Svk, which evaluates surfaces with regard to their ability to absorb and to serve fluid to adjust the lubricating properties in tribosystems, did not show any correlation with force and temperature, contrary to expectations.

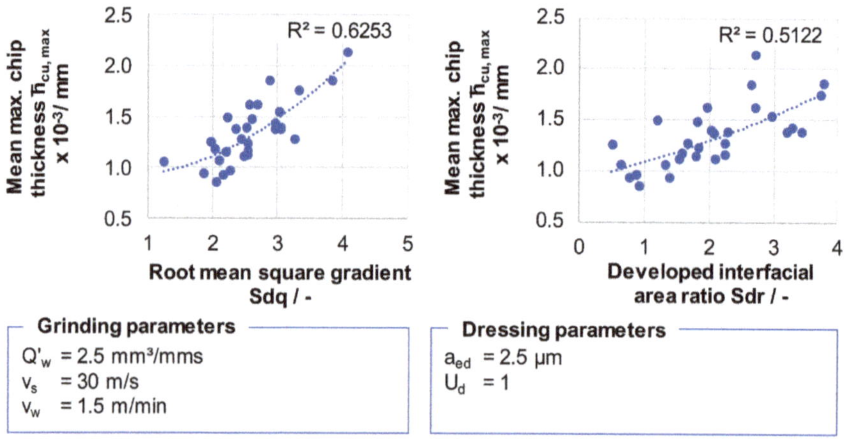

Figure 8. Correlation between the hybrid parameters and the mean max. chip thickness.

However, topography parameters that describe the characteristics of the peaks of a topography showed correlations with the thermo-mechanical stress collective. The peak extreme height Sxp is the difference of heights at the areal material ratio values $p = 0\%$ and $q = 50\%$. The corresponding material portions p and q correlated strongly with the grain protrusion. Influences on the maximum temperature T_{max} and the occurring grinding forces F_t and F_n were measured. As the grain protrusion increased and the value of Sxp raised, the grinding force and the temperature dropped. The same applied to the relationship between the reduced peak height Spk, which corresponds to the reduced height of the peaks above the core surface, and the grinding forces in the normal and tangential directions.

The peak material volume Vmp describes the material volume for an areal material ratio value of 10%. It corresponds to the volume, which is occupied by the tips of the topography. The larger the volume of the tips, the higher is the protrusion of the grains, or the amount of grains out of the core volume of the topography. If the peak material volume Vmp is small, then the core volume Vmc is correspondingly larger. This is the case if grinding wheels are smooth with a small grain protrusion. In the case of a small grain protrusion, only a small quantity of grinding fluid can enter the contact zone, which influences the contact conditions. As the peak material volume Vmp increased, the tangential grinding force F_t, as well as the maximum temperature T_{max}, dropped significantly, as presented in Figure 9. The coefficient of determination is in the range of $R^2 = 0.5$–0.6. The relatively small R^2 indicates that the parameters Sxp, Vmp, and Spk influence the thermo-mechanical stress collective, but are likewise not suitable for describing the influence of the grinding wheel topography on the thermo-mechanical stress collective solely.

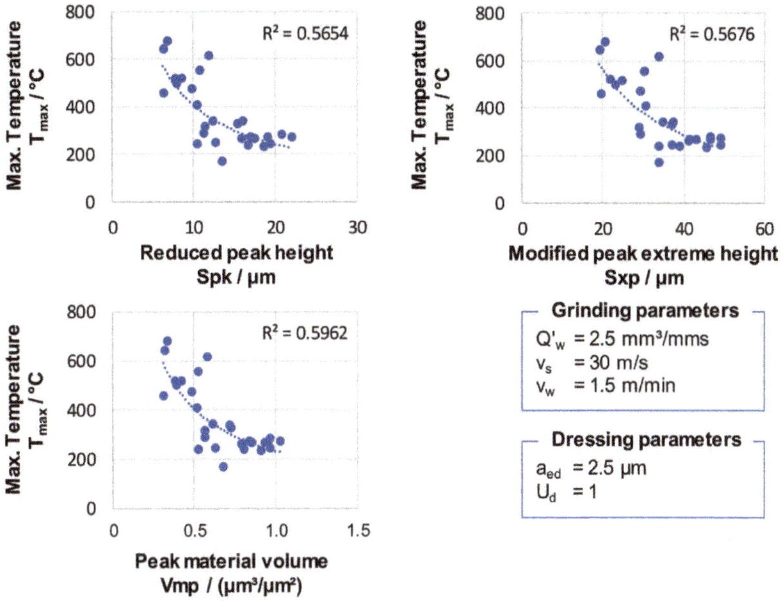

Figure 9. Correlation between volume and functional topography parameters and the process results.

The feature parameters include the peak density of the surface Spd and the arithmetic mean peak curvature Spc. The peak density Spd indicates the number of roughness peaks of a topography relative to the considered area [10]. This parameter ignores that a large part of the identified tips does not come into engagement because of shading effects during the grinding process. Accordingly, the meaningfulness of this parameter is very low. With regard to the process results, Spc showed a correlation with the maximum grinding force in normal and tangential direction F_n and F_t, as well as to the maximum temperature T_{max}. The results showed distinctive trends, but the coefficients of determination R^2 were low. So the feature parameters were not taken into account for detailed investigations. The test results showed a linear correlation between the mean tangential kinematic engagement area $\bar{A}_{kin,t}$, and the mean maximum chip thickness $\overline{h}_{cu,\,max}$. This correlation becomes already clear from geometrical considerations. A similar, but lower effect showed an increase of the total tangential kinematic engagement area $A_{kin,t}$. The relationship between the average tangential kinematic engagement area $\bar{A}_{kin,t}$ and the averaged maximum chip thickness $\overline{h}_{cu,\,max}$ is described by a linear mathematical function, with a coefficient of determination of $R^2 = 0.6463$ (see Figure 10).

The mean peak angle $\overline{\beta}$, the mean rake angle $\overline{\gamma}$ and the mean sharpening angle $\overline{\delta}$ showed a strong influence on the averaged maximum chip thickness $\overline{h}_{cu,\,max}$. The coefficient of determination for a linear relationship between $\overline{\beta}$ and $\overline{h}_{cu,\,max}$ was about $R^2 = 0.69$, which corresponds to a good regression (see Figure 11). Direct connections between the averaged maximum chip thickness $\overline{h}_{cu,\,max}$ and the grinding force, as well as the temperature were not detectable. One possible reason is that the calculation of $\overline{h}_{cu,\,max}$ solely does not take the different grain protrusions during the investigations into account.

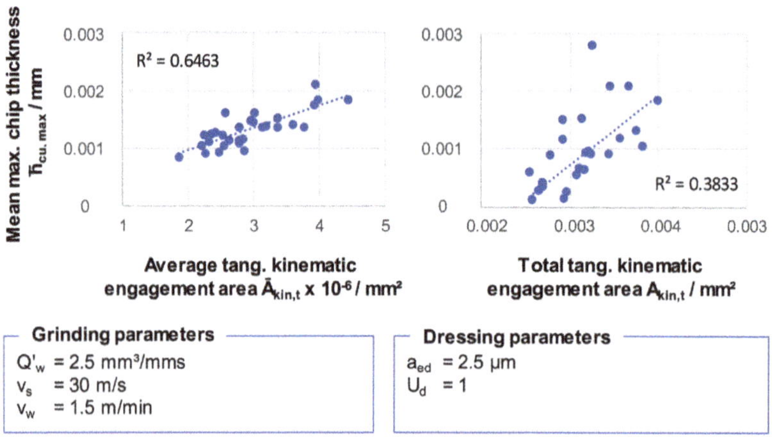

Figure 10. Correlation between kinematic engagement grain area and chip thickness.

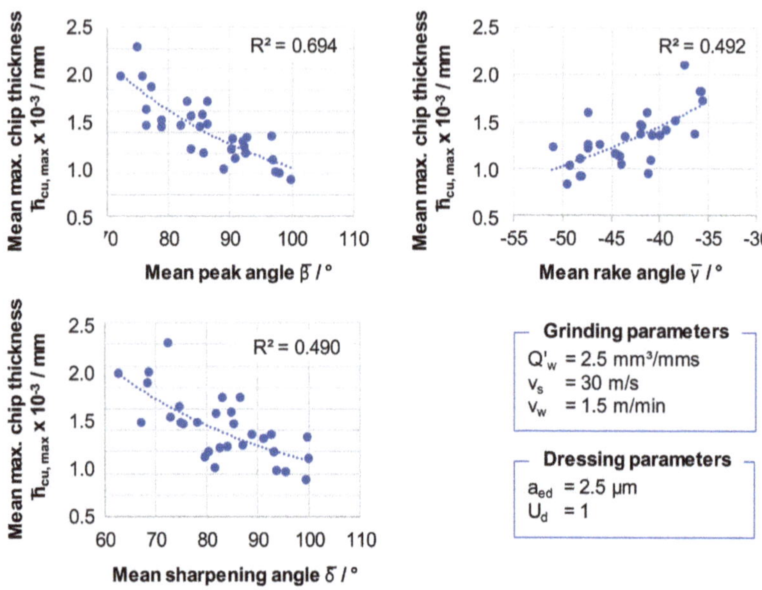

Figure 11. Correlation between kinematic engagement grain angles and the mean maximum chip thickness.

In Figure 12 the relationship between the mean rake angle $\bar{\gamma}$, the developed interfacial area ratio Sdr, and the mean tangential kinematic engagement area $\bar{A}_{kin,t}$ is shown. A strong correlation prevails between the mean sharpening angle $\bar{\delta}$ and the mean tangential kinematic engagement area $\bar{A}_{kin,t}$. An increase in the developed interfacial area ratio Sdr is associated with a rising mean tangential kinematic engagement area $\bar{A}_{kin,t}$. Based on these correlations, this means that the mean grain surface contact angle $\bar{\beta}$, $\bar{\gamma}$, and $\bar{\delta}$ as a group, and the topography characteristics Spc, Sdr, and Sdq as a group correlate as well.

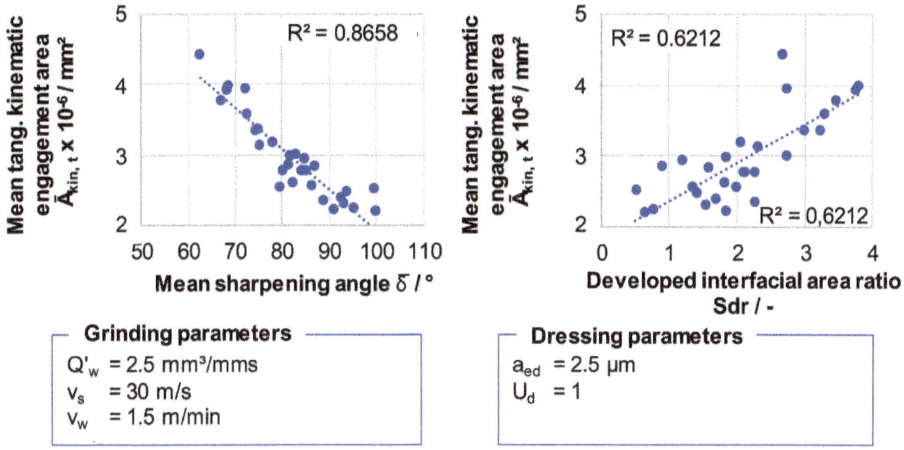

Figure 12. Correlation between the mean tangential kinematic engagement area, the mean sharpening angle and the developed interfacial area ratio.

The results of the empirical investigations showed that a change of each identified topography parameter group had an influence on the formation of the thermo-mechanical stress collective. However, the results of the investigation also allowed for the conclusion that the thermo-mechanical load in the grinding process is caused by a superposition of the influences of these identified topography parameters. Therefore, the identified topography parameters were investigated in combination.

In this context, Figure 13 shows the measured maximum grinding temperature T_{max} depending on the peak material volume Vmp and the mean peak angle $\overline{\beta}$. The results showed that the maximum grinding temperature T_{max} decreased strongly with an increasing mean peak material volume Vmp. However, the results show that topographies with a similar peak material volume Vmp did not necessarily lead to a similar maximum grinding temperature T_{max}. By a combined examination of the mean peak material volume Vmp and the mean peak angle $\overline{\beta}$ in Figure 13, it can be seen that the fluctuating maximum grinding temperatures at almost constant Vmp can be attributed to the geometry of the kinematic peak angles. If, for example, the mean peak angle $\overline{\beta}$ increased, the maximum grinding temperature T_{max} increased significantly with a constant peak material volume Vmp. This was especially noticeable at low values of the mean peak material volume Vmp. An increase of the mean peak angle from $\overline{\beta} = 75°$ to $\overline{\beta} = 95°$ at a constant mean peak material volume of Vmp = 0.4 led to an rising maximum grinding temperature T_{max} from nearly 500 °C to more than 650 °C. With increasing Vmp, the influence of the angle decreased. It can be assumed that the influence of the mean peak angle $\overline{\beta}$ decreased with increasing Vmp, since additional chip space allowed for an increased dissipation of the process heat. The dependence of the grinding temperature on the grinding wheel topography could be approximated with the help of the two parameters with a coefficient of determination of $R^2 = 0.71$ and a root mean square error of 85.2, which corresponds to a good accuracy.

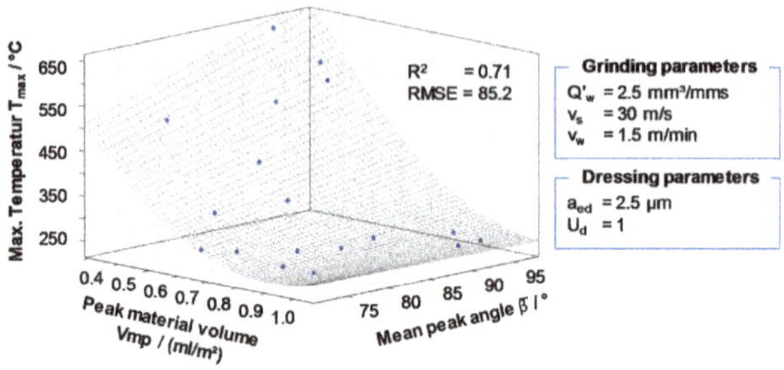

Figure 13. Grinding temperature depending on the topography parameters peak material volume and mean peak angle.

4. Discussion of the Results

Based on the presented evaluations, it can be summarized that three groups, which correlated with the temperature and force during the grinding process, with each three characteristic parameters were identified. The first group, consisting of the developed interface area ratio Sdr, the arithmetic mean peak curvature Spc, and the root mean square gradient Sdq describes the shape of the roughness on the grinding wheel surface. An increase of these parameters resulted in a reduction of the normal grinding force F_n and the maximum grinding temperature T_{max}. The second topography parameter group, consisting of the grain engagement angles mean peak angle $\overline{\beta}$, the mean sharpening angle $\overline{\delta}$, and the mean rake angle $\overline{\gamma}$, as well as the kinematic grain surface area in tangential and normal direction $A_{kin,t}$ and $A_{kin,n}$ showed a strong correlation with the mean maximum chip thickness $\overline{h}_{cu,\,max}$. In addition, they slightly correlated with the first named group. The third group of topography parameters includes functional parameters, which describe the material proportions of the topography by means of the abbott-curve; the reduced peak height Spk, the peak material volume Vmp, and the peak extreme height Sxp. These topography parameters correlated strongly with each other, but did not show any correlation with other parameters. A change of these parameters had a very strong influence on the grinding forces and the temperature. Figure 14 shows a summary of the previously identified groups, which have a strong influence the thermo-mechanical stress collective.

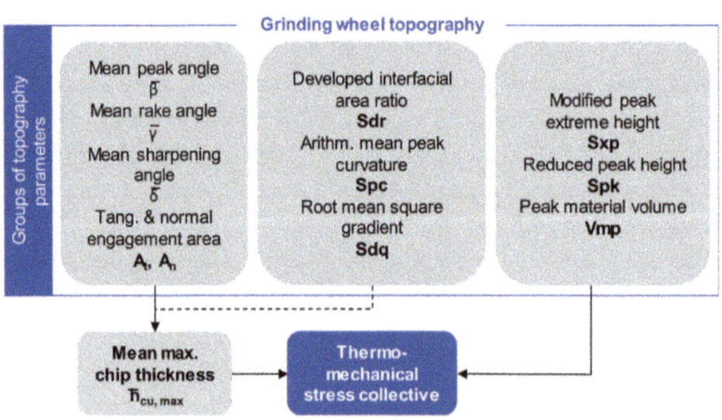

Figure 14. Overview about the topography parameter groups with an influence on the process behavior.

By a combined consideration of the influence of these three groups, the thermal loads in the grinding process were approximated with a high degree of accuracy. Thus, the maximum measured grinding temperature T_{max} could be approximated with the help of the parameters peak material volume Vmp and the mean peak angle $\overline{\beta}$ with a coefficient of determination of $R^2 = 0.71$, which corresponds to a good value, due to the high complexity in grinding technique. The research results show that by means of the identified topography parameters, it was possible to describe the influence of the topography's geometry on the thermo-mechanical stress collective based on only few geometrical parameters quantitatively. Further investigations should confirm and extend these results.

5. Conclusions

The aim of the investigations was to identify the influence of the grinding wheel topography on the thermo-mechanical stress collective during the grinding process, and to determine the topography parameters by which the influence of the topography on the thermo-mechanical stress collective can be described quantitatively. For this purpose, empirical-analytical examinations with different grinding wheels were performed. A high variation of topographies was realized by means of 13 different grinding wheel specifications and by varied dressing and sharpening strategies.

- During empirical investigations, topography parameters could be identified and classified into groups that had an influence on the thermo-mechanical stress collective
- Each group combines topography parameters, which had the same influence on the formation of the thermo-mechanical stress collective (Figure 14).
- The isolated consideration of these groups, however, did not provide a clear explanation of the dependence of the thermal and mechanical loads in the grinding process on the grinding wheel topography.
- The combined consideration of the groups around the topography parameters peak material volume Vmp and mean peak angle $\overline{\beta}$ allowed for the quantitative description of the dependence of the thermo-mechanical stress collective on the geometry of the grinding wheel topography. So, it became possible to describe the influence of the topography on the stress collective during grinding by a few topography parameters for the first time quantitatively.

In further investigations, the influence of the dressing overlap ratio, as well as the volumetric composition and the related machining volume on the grinding wheel topography and its kinematic components, as well as on the grinding process result will be investigated.

Acknowledgments: The author would like to thank the German Research Foundation (Deutsche Forschungsgemeinschaft—DFG) for the support of the depicted research within the project KL500/178-1 "Influence of the grinding wheel topography on the thermo-mechanical stress collective". Special thanks to the company TYROLIT Schleifmittelwerke Swarovski K.G., for the intensive discussion and for their support in providing specimens and grinding wheels in a variety of specifications.

Author Contributions: Sebastian Barth conceived, designed and performed the experiments, analyzed the data, interpreted the results and wrote the manuscript. Fritz Klocke contributed to the discussion and reviewed and approved the manuscript.

Conflicts of Interest: The authors declare no conflict of interest.

References

1. Schmitt, R. Abrichten von Schleifscheiben mit Diamantbestückten Rollen. Ph.D. Thesis, Technische Universität Carolo-Wilhelmina zu Braunschweig, Braunschweig, Germany, 1968.
2. Darafon, A.; Warkentin, A.; Bauer, R. Characterization of grinding wheel topography using a white chromatic sensor. *Int. J. Mach. Tools Manuf.* **2013**, *70*, 22–31. [CrossRef]
3. Frühling, R. Topographische Gestalt des Schleifscheibenschneidenraumes und der Werkstückrautiefe beim Außenrundeinstechschleifen. Ph.D. Thesis, Technische Universität Carolo-Wilhelmina zu Braunschweig, Braunschweig, Germany, 1976.
4. Paucksch, E.; Holsten, S.; Linß, M.; Tikal, F. *Zerspantechnik: Prozesse, Werkzeuge, Technologien*, 12th ed.; Studium, Vieweg+Teubner Verlag/GWV Fachverlage GmbH: Wiesbaden, Germany, 2008.
5. Uhlmann, E.; Koprowski, S.; Weingaertner, W.L.; Rolon, D.A. Modelling and Simulation of Grinding Processes with Mounted Points. *Procedia CIRP* **2016**, *46*, 599–602. [CrossRef]

6. Nowicki, B. Multiparameter representation of surface roughness. *Wear* **1985**, *102*, 161–176. [CrossRef]
7. Hecker, R.L.; Ramoneda, I.M.; Liang, S.Y. Analysis of wheel topography and grit force for grinding process modeling. *J. Manuf. Process.* **2003**, *5*, 13–23. [CrossRef]
8. Nguyen, A.T.; Butler, D.L. Correlation of grinding wheel topography and grinding performance. *J. Mater. Process. Technol.* **2008**, *208*, 14–23. [CrossRef]
9. Hübert, C.; Hahmann, D.; van der Meer, M.; Hahmann, W.C.; Pekárek, M.; Rickens, K.; Mauren, F.; Mutlugünes, Y. Charakterisierung von Schleifscheibentopographien aus fertigungstechnischer Sicht. *Diamant Hochschulwerkzeuge* **2009**, *4*, 40–47.
10. Deutsches Institut für Normung e.V. (DIN). *Geometrische Produktspezifikation (GPS)—Oberflächenbeschaffenheit;* 17.040.30 (DIN EN ISO 25178); Beuth Verlag GmbH: Berlin, Germany, 2013.
11. Duscha, M.; Klocke, F.; Wegner, H.; Gröning, H. Erfassung und Charakterisierung der Schleifscheibentopographie für die anwendungsgerechte Prozessauslegung. *Diam. Bus.* **2009**, *28*, 28–33.
12. Kassen, G. Beschreibung der Elementaren Kinematik des Schleifvorganges. Ph.D. Thesis, RWTH Aachen University, Aachen, Germany, 1969.
13. Werner, G. Kinematik und Mechanik des Schleifprozesses. Ph.D. Thesis, RWTH Aachen University, Aachen, Germany, 1971.
14. Weiß, M.; Klocke, F.; Barth, S.; Rasim, M.; Mattfeld, P. Detailed Analysis and Description of Grinding Wheel Topographies. *J. Manuf. Sci. Eng.* **2017**, *139*, 54502. [CrossRef]
15. Weiß, M. *Einfluss der Spezifikation Mehrschichtiger CBN-Schleifwerkzeuge auf das Schleifprozessverhalten: Specification Influence of Multilayer CBN-Grinding Wheels on the Grinding Process Behaviour,* 1st ed.; Apprimus Verlag: Aachen, Germany, 2016.
16. Duscha, M. Beschreibung des Eigenspannungszustandes beim Pendel- und Schnellhubschleifen. Ph.D. Thesis, RWTH Aachen University, Aachen, Germany, 2014.
17. Rasim, M. *Modellierung der Wärmeentstehung im Schleifprozess in Abhängigkeit von der Schleifscheibentopographie,* 1st ed.; Apprimus Verlag: Aachen, Germany, 2016.
18. Choi, H.-Z. Beitrag zur Ursachenanalyse der Randzonenbeeinflussung beim Schleifen. Ph.D. Thesis, Technische Universität Hannover, Hannover, Germany, 1986.
19. Batako, A.D.; Rowe, W.B.; Morgan, M.N. Temperature measurement in high efficiency deep grinding. *Int. J. Mach. Tools Manuf.* **2005**, *45*, 1231–1245. [CrossRef]
20. Demsar, J.; Curk, T.; Erjavec, A.; Gorup, C.; Hocevar, T.; Milutinovic, M.; Mozina, M.; Polajnar, M.; Toplak, M.; Staric, A.; et al. Orange: Data Mining Toolbox in Python. *J. Mach. Learn. Res.* **2013**, *14*, 2349–2353.

inventions

MDPI

Article

Influence of Cutting Speed on Subsurface Damage Morphology and Distribution in Ground Fused Silica

Georg Schnurbusch [1,*], Ekkard Brinksmeier [2] and Oltmann Riemer [1]

[1] Laboratory for Precision Machining (LFM), University of Bremen, Badgasteiner Str. 2, 28359 Bremen, Germany; oriemer@lfm.uni-bremen.de

[2] Foundation Institute of Materials Science, Badgasteiner Str. 3, 28359 Bremen, Germany; brinksmeier@iwt-bremen.de

* Correspondence: schnurbusch@lfm.uni-bremen.de; Tel.: +49-(0)421-218-51141

Received: 3 July 2017; Accepted: 4 August 2017; Published: 10 August 2017

Abstract: In optical fabrication, brittle-hard materials are used for numerous applications. Especially for high-performance optics for laser or lithography applications, a complex and consistent production chain is necessary to account for the material properties. Particularly in pre-processing, e.g., for shaping optical components, brittle material behavior is dominant which leads to a rough surface layer with cracks that reach far below the surface. This so called subsurface damage (SSD) needs to be removed in subsequent processes like polishing. Therefore, it is essential to know the extent of the SSD induced by shaping for an efficient design of precise corrective processes and for process improvement. Within this work the influence of cutting speed on SSD, in fused silica, induced by grinding has been investigated. To analyze the subsurface crack distribution and the maximum crack depth magnetorheological finishing has been appointed to polish a wedge into the ground surface. The depth profile of SSD was analyzed by image processing. For this purpose a coherent area of the polished wedge has been recorded by stitching microscopy. Taking the form deviation of the ground surface in to account to determine the actual depth beneath surface, the accuracy of the SSD-evaluation could be improved significantly. The experiments reveal a clear influence of the cutting speed on SSD, higher cutting speeds generate less SSD. Besides the influence on the maximum crack depth an influence on the crack length itself could be verified. Based on image analysis it was possible, to predict the maximum depth of cracks by means of crack length.

Keywords: optics manufacturing; precision grinding; subsurface damages; brittle hard materials

---

1. Introduction

Grinding is still an indispensable process step in the manufacturing of high precision optics, due to its high removal rates and well-controlled shaping. However, due to the prevailing material removal mechanisms, the grinding of brittle materials inevitably leads to microscopic cracks beneath the surface. This so called subsurface damage (SSD) influences the operability and lifetime of high-performance optics like semiconductor or high power laser applications [1,2]. The shaping process by grinding is followed by various corrective processes like polishing for smoothing the surface, improving the shape accuracy and removing SSD. The maximum crack depth at each process step decisively determines the extent of the polishing and corrective processes required and thus the costs of the overall optic manufacturing process [3]. According to this, a main issue in the production of high performance optics is to minimize the process-related SSD in grinding.

The undefined cutting edge and brittle material removal behavior make grinding a stochastic process. Single grains with high protrusions lead to a local overload of the normal force and cause the deepest cracks [4]. For fine-grained diamond grinding wheels the influencing variables grain protrusion and number of active grains cannot be determined directly, therefore it is nearly impossible to model the grinding process or to predict SSD. Thus, the process design for optics manufacturing still depends on empirical data. To date, the grain size serves as an orientation [5]. The stability of grinding processes as well as the accuracy in ascertaining the maximum depth of SSD is essential for increasing the efficiency in optics manufacturing. The same counts for the investigation of the influencing variables on the process-induced SSD. The accuracy of the evaluation is even more important for fine grinding where crack depths are only a few micrometers.

Nature and extent of subsurface damage in glasses makes quantitative assessment of damage difficult [6]. To date, it is still challenging to determine subsurface damage in optical glasses. Since the measurement of SSD with destructive methods is complex as well as cost-intensive and time-consuming, several attempts were undertaken to correlate process parameters and measurable output values, like surface roughness. The results of Lambropoulos et al. and Wang et al. show, however, a broad distribution in the results of these methods [7,8]. Even though there are numerous non-destructive approaches to determine the crack depths beneath the machined surface, the destructive methods still provide the most accurate results [7]. Over the past few years, the combination of magnetorheological finishing (MRF) and etching has proven to be the most appropriate strategy [7,8]. MRF provides a highly deterministic material removal process without generating new or propagating existing SSD [9]. The subsequent etching process serves to open the cracks on the polished surface to provide a better contrast in optical imaging. The damage free polishing of a wedge allows for a glance under the ground surface. Thus, not only the maximum crack depth can be determined, but also the density and shape of cracks can be observed at different depths. Resolution of this analysis obviously depends on the wedge gradient, as the area increases with decreasing wedge angles.

Simple investigations of SSD by MRF technique are limited to the determination of the maximum crack depth [10,11]. Miller et al. first analyzed the depth profile of the crack density by single micrographs along the wedge [4]. To enable analysis for a larger wedge area, adjacent microscope images were stitched by using a computer-based microscope system with a driven stage [8]. This method offers the great advantage, to analyze the recorded data by image processing. The SSD depth is calculated from the difference in height between the profiles of the ground surface and the wedge surface. In previous publications, a perfect initial surface was assumed, however, even fine grinding generates form deviations of a few micrometers. For rough cuts and pre-shaping this seems negligible, but in the case of fine grinding these deviations can be on the same scale as the SSD.

Within this work, the influence of cutting speeds on SSD for ground fused silica was investigated. To analyze the maximum crack depth and crack distribution MRF technique was used in conjunction with dry etching. Finally, for the first time, the initial surface profile was considered to enhance the accuracy to determine the actual crack depth.

2. Materials and Methods

Fused silica samples with a diameter of 100 mm were ground on an ultraprecision 5-axis grinding machine (Nanotech 500 Freeform Generator, Moore Nanotechnology Systems). The experiments comprise a variation of cutting speeds between 30 and 62 m/s, in five subdivisions, with three repetitions (Table 1). For process monitoring, the grinding forces were measured by a multicomponent dynamometer (CompactDyn, Kistler, Winterthur, Switzerland) mounted beneath the hydrostatic grinding spindle. The grinding kinematics is shown in Figure 1.

A resin bonded diamond grinding wheel with a diameter of 100 mm, D20 (C70) was applied. All samples were ground with a constant rotary worktable speed n_W = 50 rpm and radial feed v_{fr} = 10 mm/s. Based on a fine grinding process, the material removal was conducted in four passes with 20 μm and one final pass with 10 μm depth of cut. In order to guarantee a uniform initial state, all samples were pre-ground to plane shape before running the actual experiments (pre-processing, a_e = 100 μm; four passes with 25 μm). Furthermore, the grinding wheel was dressed and balanced before each grinding experiment. This ensures that the measured SSD results from the actual process parameters.

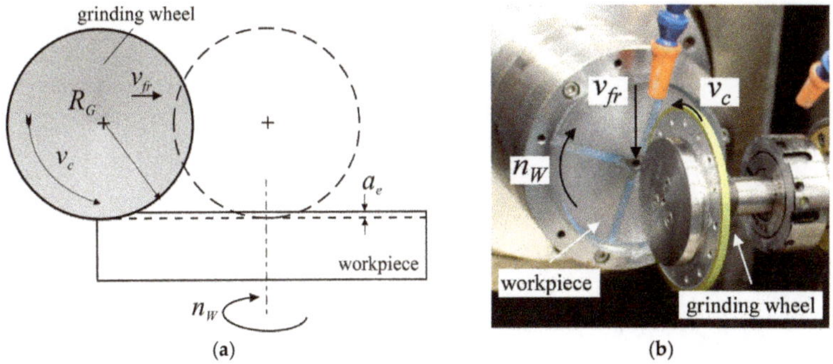

Figure 1. (**a**) Grinding kinematics—cross grinding with rotating workpiece; (**b**) Machine setup.

Table 1. Process parameters.

Process Step	Cutting Speed v_c (m/s)	Depth of Cut a_e (μm)	Worktable Speed n_W (rpm)	Radial Feed v_{fr} (mm/min)
1st step: pre-processing	30	100 (4 × 25)	50	10
2nd step: SSD experiments	30–62	90 (4 × 20, 1 × 10)	50	10

3. Results

3.1. Grinding Force and Surface Roughness

Based on the constant rotary worktable speed, the feed rate decreases continuously from edge to center. Since the remaining parameters are kept constant, this leads to a reduction in material removal rate and thus a reduction in chip thickness. Consequently, normal forces (Figure 2a) and roughness decrease from edge to center. In conventional grinding, it is generally assumed that higher cutting speeds lead to a lower chip thickness and thus also to lower grinding forces [12]. In the experiments presented in this work, however, the opposite is shown. Larger cutting speeds led to higher normal forces (Figure 2b). For brittle materials like fused silica, the removal mechanisms depend on the chip thickness. Due to a higher rate of ductile material removal, the forces increase with the cutting speed [13]. To make the results comparable and regarding the position of the maximum crack depth c_{max} at the MRF wedge, the presented values were obtained at a distance of 15 mm from workpiece center.

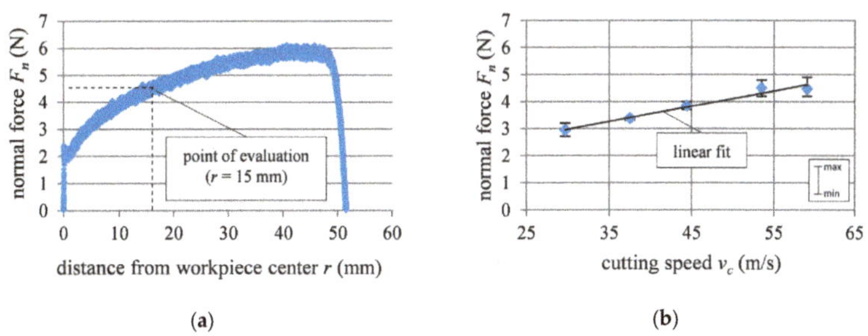

Figure 2. (**a**) Typical force curve (F_n) at v_c = 62 m/s for constant worktable speed n_W; (**b**) Normal forces for different cutting speeds (distance from workpiece center 15 mm).

The surface roughness was measured by white light interferometry (WLI, Talysurf CCI HD, Taylor Hobson) with 20× magnification. Along circles of different radii six spots with an area of 840 × 840 μm² were measured with an angular distance of 60°. To exclude the waviness of the surface a long pass filter was set to 250 μm.

The results of the roughness measurement (e.g., in Figure 3) show an opposite trend towards normal force. This agrees well with the assumption that the proportion of ductile material behavior increases with higher cutting speeds.

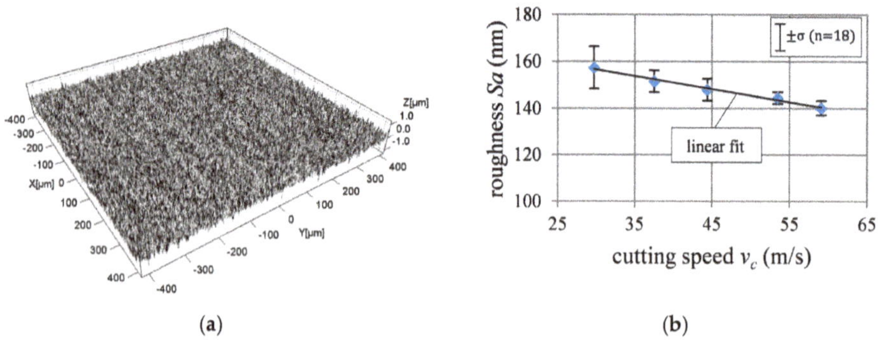

Figure 3. (**a**) White light interferometer (WLI) measurement (20× magnification): 2D-surface roughness (v_c = 30 m/s, *Sa* = 160 nm); (**b**) Surface roughness *Sa* for different cutting speeds (distance from workpiece center 15 mm) with linear fit.

3.2. *Subsurface Damage Evaluation*

For the evaluation of subsurface damage, the micrographs of the MRF wedge need to be merged with the depth profile. In order to maintain a high accuracy, the samples were provided with a fiducial mark by laser engraving after grinding. It is thus possible to align the samples for different measurements and to link the data of tactile measurements and microscope images. Figure 4 shows the wedge geometry and the entire evaluation procedure.

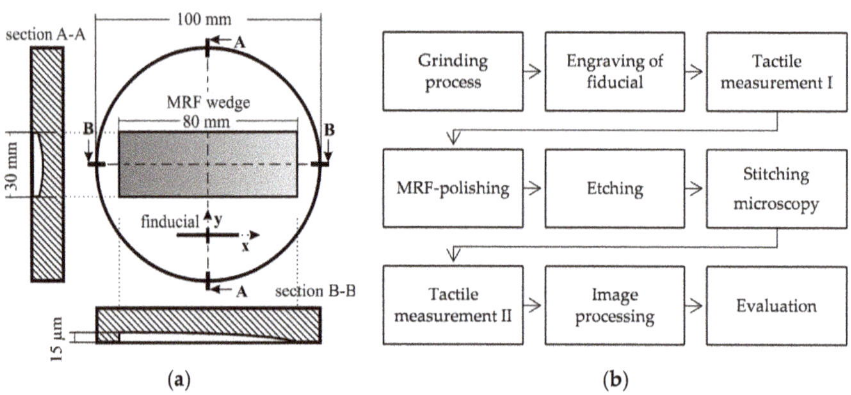

Figure 4. (**a**) Position and geometry of the magnetorheological finishing (MRF) wedge; (**b**) Sample processing chain.

After MRF preparation, the wedge surfaces were dry etched. Compared to wet etching the material removal by dry etching can be controlled with a sub-micron precision [14]. The micrographs of the wedge surface were taken with a ZEISS Axio Imager.Z2 Vario with 500× magnification. 750 micrographs were stitched together to an interrelated area of 40 × 1 mm^2. To minimize contamination from particles the samples were cleaned in an ultrasonic bath and microscopy was carried out under clean room conditions (ISO 7). In order to determine the actual depth of the wedge as accurately as possible, the distance between the ground surface (tactile measurement I) and the wedge surface (tactile measurement II) was calculated from the two tactile measurements.

Due to the form deviation, differences between the polished wedge surface and the actual depth profile of up to 2 μm (Figure 5) occur in the individual experiments. This is particularly relevant for fine grinding processes with low SSD. Within the experiments, the differences correspond to 30–40% of the measured maximum crack depth.

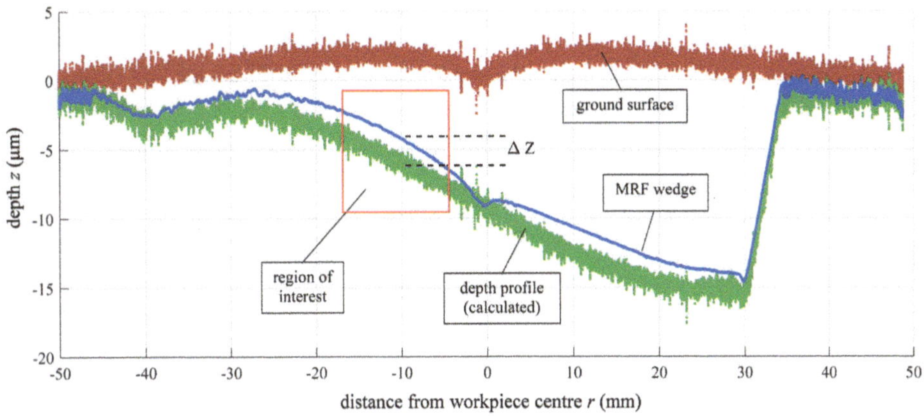

Figure 5. Tactile measurement of ground surface, MRF wedge and depth profile.

The number of cracks decreases with depth along the MRF wedge. With increasing depth of the wedge, more and more cracks are removed by polishing until the last fracture event can be determined as the maximum crack depth in the micrographs (Figure 6).

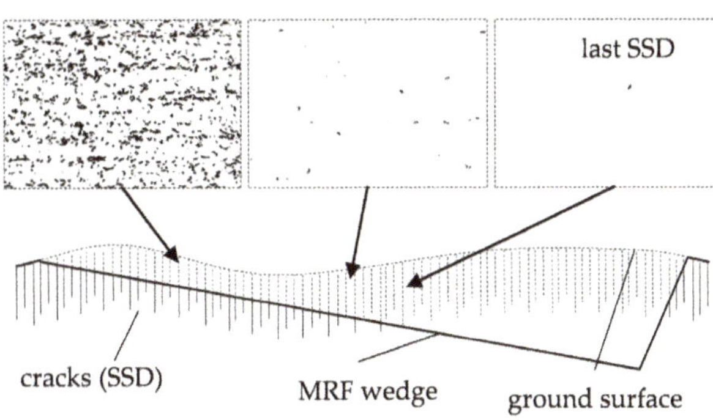

Figure 6. Sketch of subsurface damage (SSD) appearance along the MRF wedge.

To determine the maximum crack depth c_{max} the lateral position of the last defect on the micrograph of the polished and etched wedge surface is merged together with the corresponding information of the calculated depth profile (Figure 7).

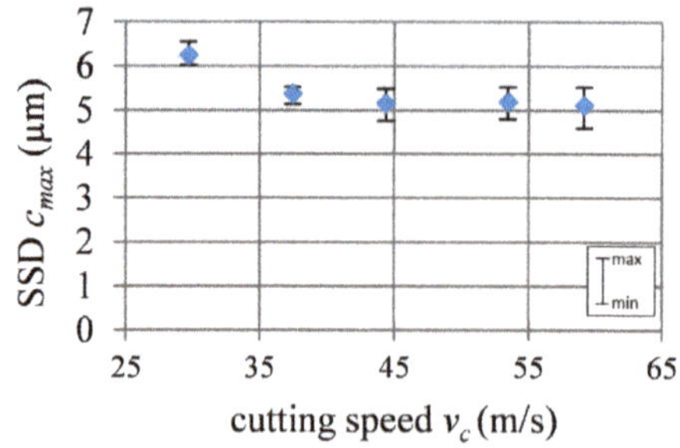

Figure 7. Maximum crack depth depending on the cutting speed.

3.3. SSD Analysis by Image Processing

Besides the evaluation of the maximum crack depth, the micrographs of the MRF wedge provide the opportunity to examine the crack distribution with increasing depth. Thus, a more extended examination on the influence of process parameters on grinding results is possible. The high number of cracks resulting from the size of the inspected wedge area requires a software-supported evaluation. Therefore the open source software, ImageJ, has been utilized. The software is a highly popular tool in biology in a wide range of applications and provides a segmentation algorithm. This enables the analysis of the position and the geometrically relevant data for each individual crack on the stitched microscope image. After binarization of the images, the cracks appear as uniform black lines on a white background. To specify the crack length, a best fit ellipse is applied to every single crack (Figure 8).

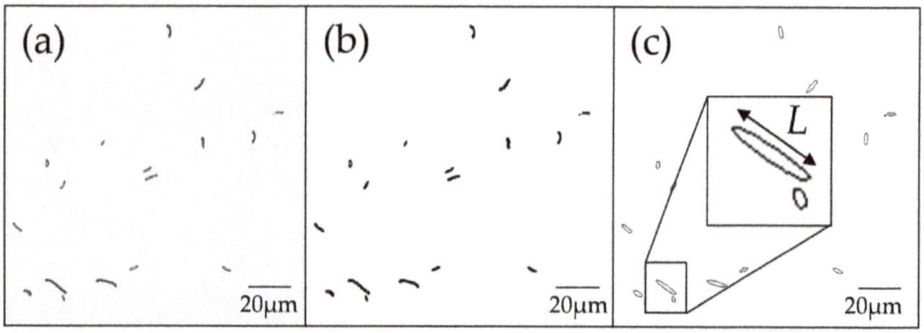

Figure 8. SSD-Image Processing (ImageJ): (**a**) raw microscopy data; (**b**) binary conversion; (**c**) determination of crack length L by best fit ellipse.

On average, 150,000 cracks were evaluated for each sample with this method. Figure 9 shows the ratio of the number of cracks of different lengths to the total crack number (relative frequency) for different cutting speeds.

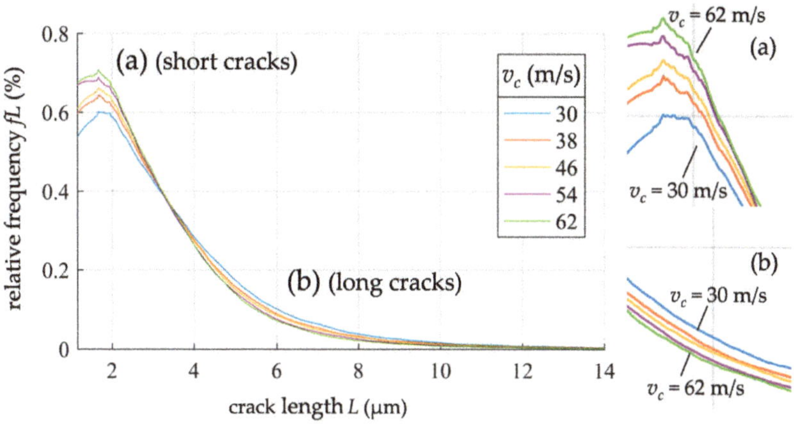

Figure 9. Relative frequencies of the crack length L (filtered data).

After smoothing by a Savitzky–Golay filter, it is easy to see that the distributions differ. As the cutting speed increases, the amount of shorter cracks (Figure 9a) increases and the amount of longer cracks decreases (Figure 9b).

4. Discussion

4.1. Maximum Crack Depth

Typically, the crack density decreases along the wedge, i.e., with increasing depth, following a power law function [15]. At the scale of the surface roughness, the individual cracks intersect and Suratwala et al. describe this as the rubble zone. A few microns below the surface, the cracks can be detected individually. Within our experiments, with the given MRF wedge geometry, the lateral distance between the last cracks could be up to 500 μm. This underlines the importance of size and the slope of the polished wedge.

After an initial decrease in the maximum crack depth with an increasing cutting speed, a seemingly constant depth is obtained above 45 m/s (Figure 7). Besides chip thickness reduction, higher spindle speeds and resulting higher cutting speeds lead to stronger vibrations [16]. This may explain why there is no further improvement of the maximum SSD-depth at cutting speeds higher than 45 m/s. The single-grain load, as the main cause for SSD, seems to be confirmed. In theory, the influence of the material removal rate and thus process parameters such as cutting speed, feed rate and infeed can be easily assessed. In practice, however, the stochastic nature of the grinding wheel as well as other influences such as limited machine rigidity have influence on the maximum crack depth.

Miller et al. have shown that the addition of a small amount of larger grains to a fine slurry for polishing leads to a significant increase in the measured crack depth [4]. The normal force is distributed to a smaller number of particles than in a homogeneous particle distribution and the maximum individual grain loads are increased. In the grinding process, the effective grain size corresponds to the individual grain protrusion, which directly depends on the grain size.

The individually polished and etched wedge surfaces differ considerably, in crack distribution. Figure 10 shows samples with low and high maximum crack depths. The sample with low SSD shows a rather arbitrarily arranged crack pattern (Figure 10a), whereas in case of significantly larger SSD (Figure 10b), the ordered crack formation is particularly remarkable. Large protrusion of individual grains leads to higher individual loads. As a result, it is not possible to assess the process results by means of maximum SSD only in conjunction with the cutting speed e.g., vibrations or variation in grain protrusion.

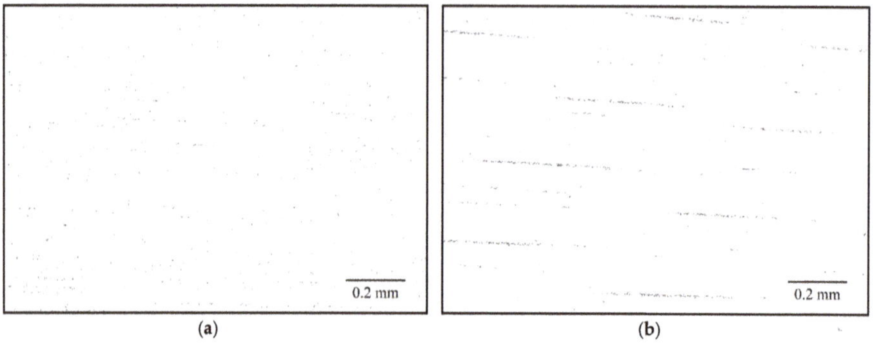

Figure 10. Micrograph of the MRF wedge surface after etching: (**a**) sample with low SSD (v_c = 54 m/s; c_{max} = 4.6 μm); (**b**) sample with larger SSD (v_c = 30 m/s; c_{max} = 6.8 μm).

As stated above, the maximum crack depth c_{max} results from one particular or only a few grains. It is thereby subject to a certain variation by the nature of the microstructure of the grinding wheel and the grinding process. Both, crack depth and crack length, depend directly on the grain load. The maximum crack depth c_{max} should therefore also lead to the maximum crack length L_{max}. Within the experiments L_{max} shows a linear decay with depth. Figure 11 shows L_{max} at intervals of 50 μm along the polished wedge. To avoid the influence of outliers by intersecting cracks on the linear fit, the double median crack length L_{Mdn} is set as cut-off criteria.

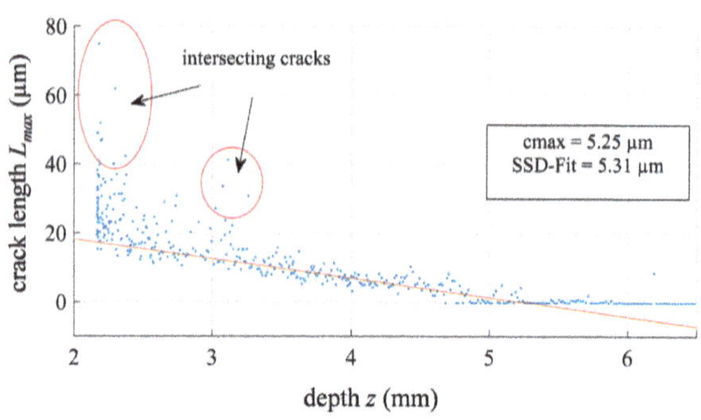

Figure 11. Maximum crack lengths L_{max} as a function of depth (v_c = 54 m/s) with linear fit.

The zero point of the fit provides a very good correlation with the measured maximum crack depths (Figure 12).

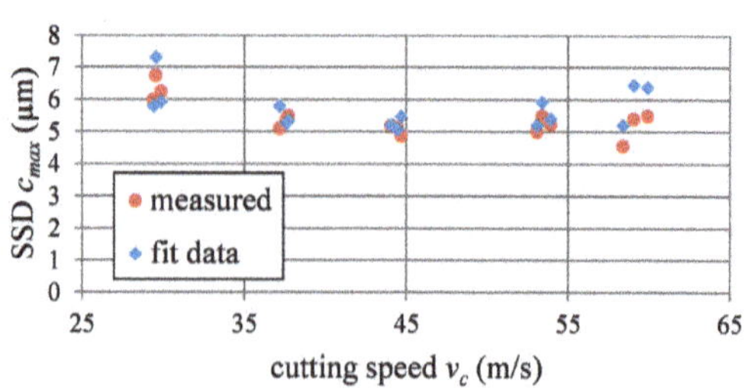

Figure 12. Measured and calculated maximum crack depths.

4.2. Fracture Mechanics

Determination of the exact surface topography of a grinding wheel is difficult, even for coarse-grained tools. Furthermore, the surface changes continuously by wear or is completely reshaped by dressing. Therefore, attempts are often made to simplify the grinding process by single grain scratch behavior. The basis for most of the considerations is the work of Lawn, who investigated the cracking mechanisms of brittle materials under indentation [17]. His experiments showed the formation of median and lateral cracks as a function of loading and unloading on the workpiece surface. He showed that the maximum radial crack length c_r correlates with the load on the indenter, i.e., $P{\sim}{c_r}^{3/2}$[18].

$$c_r = \left(\frac{\chi_r P}{K_{Ic}}\right)^{2/3} \text{with } \chi_r \sim E/H \tag{1}$$

With the indentation constant χ_r (0.046 for fused silica [19]), indenter load P, fracture toughness K_{Ic}, hardness H and Young's modulus E.

Due to the stochastic nature of the microtopography of grinding wheels, loads on a single grain can not be determined. Equation 1, nevertheless, reveals the maximum grain load to be calculated from the maximum radial crack length c_r, which corresponds to the maximum crack depth c_{max}.

$$P = \frac{K_{Ic}\, {c_{max}}^{3/2}}{\chi_r} \tag{2}$$

Using equation 2 and the measured maximum crack depths (Figure 7), the maximum individual grain loads P are about 0.2 N for the grinding wheel applied in this work. This corresponds to approximately 5% of the measured average normal force. In addition to the evaluation of the maximum crack depth, the wedge method also makes it possible to determine the crack length L on the polished surface. Based on the Hertzian indentation mechanics Miller et al. found a correlation between average crack length and max crack depth [4].

Assuming that the diamond grains are idealized spherical abrasives the circles of contact are defined by the Hertzian contact zone, with the contact circle radius a.

$$a = \left(\frac{4}{3}\frac{\mathrm{k}}{\mathrm{E}} P\mathrm{r}\right)^{1/3} \tag{3}$$

With the elastic mismatch factor k, the applied load P and the radius of the abrasive r. The factor k relates to the different Young's moduli and Poisson's ratio ν of the workpiece and indenter material, where the primes refer to the indenter.

$$\mathrm{k} = \frac{9}{16}\left[\left(1-\nu^2\right) + \frac{\mathrm{E}}{\mathrm{E}'}\left(1-\nu'^2\right)\right] \tag{4}$$

In contrast to static indentation, a sliding indenter causes an asymmetry in the tensile stress field. Since the tensile stresses are relatively small near the front end of the indenter the likelihood of cone cracks increases in this region. This also predicts that the crack may not completely encircle the entire contact radius [20]. Miller et al. assumed that the crack propagates over only a quarter of the contact circumference.

$$L \approx \frac{2\pi a}{4} = \frac{\pi}{2}\left(\frac{4}{3}\frac{\mathrm{k}}{\mathrm{E}'} P\mathrm{r}\right)^{1/3} \tag{5}$$

Finally, the calculated crack length L_{calc} as a function of c_{max} according to Equations (2) and (5) shows appropriate accordance with the determined average crack length L_{Mdn} from the optical evaluation (Figure 13).

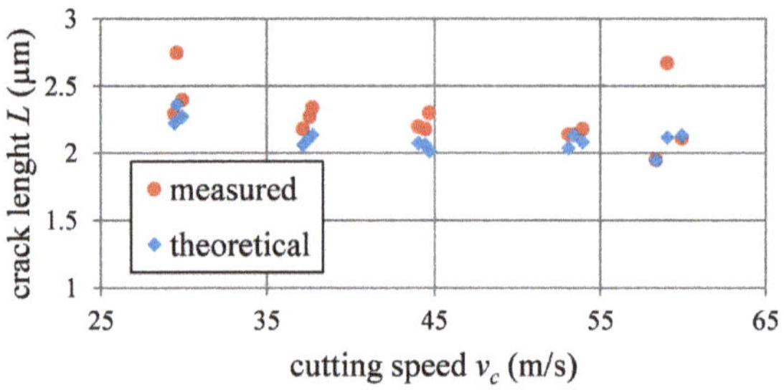

Figure 13. Measured and calculated crack length for different cutting speeds.

5. Conclusions

To date there is no alternative to the destructive techniques for the evaluation of SSD in optical glasses. By using modern stitching microscopes and image processing, the MRF wedge method offers the highest accuracy in the determination of crack distribution and maximum crack depth. Furthermore, a thorough investigation of the removal mechanisms of the grinding process of optical glasses is enabled.

Within this work subsurface damage (SSD) was directly determined by the MRF wedge technique for a series of fused silica ground with different cutting speeds from 30 to 62 m/s. To improve the accuracy of the SSD evaluation method wet etching was replaced by dry etching. Furthermore, the form deviation of the ground surfaces was taken into account to determine the actual crack depth. Thus, a significant influence of the cutting speed on SSD for a fine grinding process, as it is found in process chains of optical fabrication, could be demonstrated. With stitching microscopy and image processing the gapless evaluation of the density of cracks and the distributions of crack lengths along the wedge surface has been implemented. Thus, a dependency of the crack length distribution on the cutting speed could be confirmed. Based on the model of Miller [4], the relation between calculated crack length L_{calc} as a function of c_{max} and median crack length L_{Mdn} could be demonstrated for SSD of only a few microns. Furthermore, a relationship between the maximum crack length as a function of depth and the maximum crack depth could be found. This and the different crack patterns of the SSD samples supports the thesis that the maximum crack depth is caused by a few individual grains that also produce the largest crack length.

Author Contributions: Georg Schnurbusch, Ekkard Brinksmeier and Oltmann Riemer designed the experiments; Schnurbusch performed the experiments and materials investigation; Georg Schnurbusch and Oltmann Riemer analyzed the data. Ekkard Brinksmeier contributed the experimental infrastructure; Georg Schnurbusch, Ekkard Brinksmeier and Oltmann Riemer discussed the results and wrote the paper.

Conflicts of Interest: The authors declare no conflict of interest.

Abbreviations

a	Hertzian contact circle radius	L_{Mdn}	average crack length (median)
a_e	depth of cut	MRF	magnetorheological finishing
c_{max}	maximum crack depth	ν	Poisson's ratio (workpiece)
c_r	radial crack length	ν'	Poisson's ratio (indenter)
E	Young's Modulus (workpiece)	n_W	worktable speed
E'	Young's Modulus (indenter)	P	indenter load
f_L	relative frequency	R	grain size (radius of the abrasive)
F_n	normal force	Sa	average surface area roughness
H	hardness	SSD	subsurface damage
K	mismatch factor	v_c	cutting speed

K_{Ic}	fracture toughness	v_{fr}	radial feed
L	crack length	WLI	white light interferometer
L_{calc}	calculated crack length	χ_r	indentation constant
L_{max}	maximum crack length		

References

1. Hed, P.P.; Edwards, D.F.; Davis, J.B. Subsurface damage in optical materials: Origin, measurement & removal. In Proceedings of the Optical fabrication and testing workshop, Santa Clara, CA, USA, 2 November 1988.
2. Neauport, J.; Ambard, C.; Bercegol, H.; Cahuc, O.; Champreux, J.P.; Charles, J.L.; Cormont, P.; Darbois, N.; Darnis, P.; Destribats, J.; et al. Optimizing fused silica polishing processes for 351 nm high power laser application. In *Laser-Inducted Damage in Optical Materials, Proceedings of the SPIE, Volume 7132, Boulder, CO, USA, 22 September 2008*; Exarhos, G.J., Ristau, D., Eds.; SPIE: Bellingham, WA, USA, 2008. [CrossRef]
3. Comley, P.; Morantz, P.; Shore, P.; Tonnellier, X. Grinding metre-scale mirror segments for the E-ELT ground based telescope. *CIRP Annals-Manufacturing Technology* **2011**, *60*, 379–382. [CrossRef]
4. Miller, P.E.; Suratwala, T.I.; Wong, L.L.; Feit, M.D.; Menapace, J.A.; Davis, P.J.; Steele, R.A. The distribution of subsurface damage in fused silica. In *Laser-Inducted Damage in Optical Materials, Proceedings of SPIE Volume 5991, Boulder, CO, USA, 19 September 2005*; Exarhos, G.J., Guenther, A.H., Eds.; SPIE: Bellingham, WA, USA, 2006. [CrossRef]
5. Lambropoulos, J.C. From Abrasive Size to Subsurface Damage in Grinding. *TOPS* **2000**, *42*, 17–18. [CrossRef]
6. Lucca, D.A.; Brinksmeier, E.; Goch, G. Progress in Assessing Surface and Subsurface Integrity. *CIRP Ann.* **1998**, *47*, 669–693. [CrossRef]
7. Wang, L.; Li, Y.; Han, J.; Xu, Q.; Guo, Y. Evaluating subsurface damage in optical glasses. *JEOS:RP 6* **2011**, *6*, 1–16. [CrossRef]
8. Menapace, J.A.; Davis, P.J.; Steele, W.A.; Wong, L.L.; Suratwala, T.I.; Miller, P.E. MRF Applications: Measurement of process-dependent subsurface damage in optical materials using the MRF wedge technique. In *Laser-Inducted Damage in Optical Materials, Proceedings of SPIE Volume 5991, Boulder, CO, USA, 19 September 2005*; Exarhos, G.J., Guenther, A.H., Eds.; SPIE: Bellingham, WA, USA, 2006. [CrossRef]
9. Menapace, J.A.; Davis, P.J.; Steele, W.A.; Wong, L.L.; Suratwala, T.I.; Miller, P.E. Utilization of magnetorheological finishing as a diagnostic tool for investigating the three-dimensional structure of fractures in fused silica. In *Laser-Inducted Damage in Optical Materials, Proceedings of SPIE Volume 5991, Boulder, CO, USA, 19 September 2005*; Exarhos, G.J., Guenther, A.H., Eds.; SPIE: Bellingham, WA, USA, 2006. [CrossRef]
10. Randi, J.A.; Lambropoulos, J.C.; Jacobs, S.D. Subsurface damage in some single crystalline optical materials. *Appl. Opt.* **2005**, *44*, 2241–2249. [CrossRef] [PubMed]
11. Li, Y.; Zhenga, N.; Li, H.; Houa, J.; Lei, X.; Chena, X.; Yuana, Z.; Guoa, Z.; Wanga, J.; Guob, Y.; Xua, Q. Morphology and distribution of subsurface damage in optical fused silica parts: Bound-abrasive grinding. *Appl. Surf. Sci.* **2011**, *257*, 2066–2073. [CrossRef]
12. Klocke, F. *Manufacturing Processes 2: Grinding, Honing, Lapping*; Springer: Heidelberg, Germany, 2009; pp. 251–269.
13. Wang, H.; Subhash, G.; Chandra, A. Characteristics of single-grit rotating scratch with a conical tool on pure titanium. *Wear* **2001**, *249*, 566–581. [CrossRef]
14. Bollinger, L.D.; Gallatin, G.M.; Samuels, J.; Steinberg, G.; Zarowin, C.B. Rapid, non-contact optical figuring of aspheric surfaces with Plasma Assisted Chemical Etching (PACE). In *Advanced Optical Manufacturing and Testing, Proceedings of SPIE 1333, San Diego, CA, USA, 1 July 1990*; Sanger, G.M., Reid, P.B., Eds.; SPIE: Bellingham, WA, USA, 1990. [CrossRef]
15. Suratwala, T.I.; Wong, L.L.; Miller, P.E.; Feit, M.D.; Menapace, J.A.; Steele, R.A.; Davis, P.J.; Walmer, D. Subsurface mechanical damage distributions during grinding of fused silica. *J. Non.-Cryst. Solids* **2006**, *352*, 5601–5617. [CrossRef]
16. Huang, H.; Yin, L. High speed grinding performance and material removal mechanism of silicon nitride. *Initiat. Precis. Eng. Begin. Millenn.* **2002**, 416–420. [CrossRef]
17. Lawn, B.R.; Swain, M.V. Microfracture beneath point indentations in brittle solids. *J. Mater. Sci.* **1975**, *10*, 113–122. [CrossRef]

18. Lawn, B.R.; Evans, A.G.; Marshall, D.B. Elastic/plastic indentation damage in ceramics: The mediad/radial crack system. *J. Am. Ceram. Soc.* **1980**, *63*, 574–581. [CrossRef]
19. Anstis, G.R.; Chantikul, P.; Lawn, B.R.; Marshall, D.B. A critical evaluation of indentation techniques for measuring fracture toughness: I, direct crack measurements. *J. Am. Ceram. Soc.* **1981**, *64*, 533–538. [CrossRef]
20. Lawn, B.R. Partial cone crack formation in a brittle material loaded with a sliding spherical indenter. *Proc. R Soc. Lond. Ser. A* **1967**, *299*, 307–316. [CrossRef]

inventions

MDPI

Article

Stochastic Kinematic Process Model with an Implemented Wear Model for High Feed Dry Grinding

Michal Kuffa *, Fredy Kuster and Konrad Wegener

Institute of Machine Tools and Manufacturing, ETH Zürich, Leonhardstrasse 21, 8092 Zürich, Switzerland; kuster@iwf.mavt.ethz.ch (F.K.); wegener@iwf.mavt.ethz.ch (K.W.)

* Correspondence: kuffa@iwf.mavt.ethz.ch; Tel.: +41-44-632-78-81

Received: 12 October 2017; Accepted: 14 November 2017; Published: 16 November 2017

Abstract: This paper considers heavy duty grinding with resin bonded corundum grinding wheels and without lubrication and cooling. A vertical turning machine redesigned to a grinding machine test bench with a power controlled grinding spindle is used in all of the experiments, allowing high tangential table feed rates up to 480 m/min. This special test-rig emulates the railway grinding usually done by a railway grinding train. The main test-rig components are presented and the resulting kinematics of the experimental set-up is described. A stochastic kinematic grinding model is presented. A wear model that is based on the kinematic description of the grinding process is set up. Grain breakage is identified as the main wear phenomenon, initiated by grain flattening and micro-splintering. The wear model is implemented into the stochastic kinematic modelling. The workpiece material side flow and spring back are considered. The simulation results are validated experimentally. The workpiece surface roughness is compared and a good agreement between simulation and experiment can be found, where the deviation between the experiment and the simulation is less than 15% for single-sided contact between the grinding wheel and the workpiece. Higher deviations between simulation and experiment, up to 24%, for double-sided contact is observed.

Keywords: wear modelling; self-sharpening; high-performance dry grinding; surface roughness

1. Introduction

Three important global trends of manufacturing technology in recent years were observed: moving towards increasing performance in material removal rate, higher resulting surface quality, and the reduction of coolant consumption. Chip transport from the cutting zone, cooling and lubrication are the main tasks of the coolant. Environmental impact, costs, swarf, recycling, and negative effects on the machine operator stand in contrast to the previously mentioned advantages. Grinding is often used for finishing applications to generate smooth surfaces with high integrity and low form deviations, but grinding is also yet increasingly challenging other manufacturing technologies such as high-speed or high-performance milling and turning. The implementation of high performance machining processes in general have become increasingly significant among cutting processes and could already modify many industrial manufacturing process chains by substituting one or more machining steps, and thus save cost.

As stated by Bhaduri and Chattopadhyay [1], but also by Kopac and Krajnik [2], high quality parts require high-performance grinding technologies when considering not only superior surface finish and high removal rate. The literature classifies high performance grinding not clearly distinguishable from conventional grinding and covers different grinding techniques such as: High Speed Grinding (HSG), High Efficiency Deep Grinding (HEDG), Speed-Stroke Grinding, High-Performance Surface Peel

Grinding, High Work Speed Grinding, High Efficiency Grinding, Heavy Duty Grinding, and Creep Feed Grinding (CFG) [2–11]. All of these grinding techniques are used with metalworking fluids contrasting to the dry grinding process presented here.

Linke [12] gave in her review a broad overview on the recent grinding wheel wear examinations and modelling and classified four main wear mechanisms: grain surface layer wear, grain splintering, grain-bond-interface wear, and bond wear. Linke [12] distinguished between micro wear where cutting edge sharpness is lost and macro wear as loss of tool profile.

Jiang et al. [13] introduced a three-dimensional (3D) model predicting the surface roughness when considering wear and dressing effects indicating dressing as the most influential parameter. A kinematic simulation predicting the surface roughness from Liu et al. [14] modelled three different abrasive grain shapes and ductile and brittle fracture components by a single point diamond dressing tool. Jiang et al. [13,14] judged dressing to be the most influential parameter on the predicted surface roughness and grain shapes do not affect the resulting surface roughness. A stochastic process model of abrasive wear, which is based on the Poisson distribution of the wear phenomenon, was presented by Kacalak, Kasprzyk and Krzyżyński [15]. Meng and Ludema [16] presented 28 different erosion wear equations from the literature having surveyed 5466 different papers. The synthesis of those appears impossible to them.

In many dry grinding applications, resin bonded grinding wheels are used due to their self-sharpening ability, cool cut, and damping capability. In this work, the high-performance dry grinding process for surface face grinding with resin bonded corundum grinding cup wheels is analysed at high feed speeds. Especially, the self-sharpening effect is analysed. An analytical wear model considering grain failure phenomena is introduced.

2. Experimental Methods

The same test-rig as described in [17] is used. The following equipment is used for all experiments: vertical lathe SEDIN 1525 with 20 kW grinding spindle typical in rail grinding, 3D-dynamometer Kistler 9366CC, roughness measuring system Taylor Hobson Talysurf, profile, and waviness measurement sensor Micro Epsilon ILD 2300-10LL are used.

A detailed view of the experimental set-up is given in Figure 1. The marked tilt angle is adjusted with a hydraulic cylinder and can be positioned in a positive and negative direction within ±15°. The grinding wheel is clamped from outside with one fixed and two movable chuck jaws. The grinding spindle is connected directly with the cover plate of the Kistler force measuring platform. This guarantees the coordinate system conforms with force measurements without the need of coordinate transformation. The axial movement of the grinding spindle and the adjustment of depth of cut a_p during the process are performed with the pneumatic cylinder mounted behind the spindle. Two pillars guide the traverse with the spindle. The exact adjustment of the depth of cut is not possible due to the power controlled spindle support that is used in railway grinding. Force control is typical also for other high-performance dry grinding applications. The target current for the electric engine is set and the pressure is continuously adjusted during the process in the pneumatic cylinder. The tilt angle is adjusted with the hydraulic cylinder and the grinding spindle is guided on two milled slides. The ring-like workpiece is placed on the rotary table that is fixed by four chuck jaws. The whole grinding spindle construction is mounted in a frame. This frame guarantees the required stiffness and strength for the process forces and torques. A laser sensor mounted on the right-hand column is used for the surface measurements.

The cutting speed during all of the experiments is kept constant at 47 m/s at the outer circumference of the grinding wheel and the feed rate of the workpiece is 120 m/min. A resin bonded corundum grinding wheel with comparatively coarse mesh 20 is applied in the experiments. All of the experiments are carried out without metalworking fluid. Workpiece material used is 58CrMoV4 with hardness of 26 HRC. A 6 mm wide facet is ground in all of the experiments performed, however the workpiece could be turned prior to any experiment to the desired width.

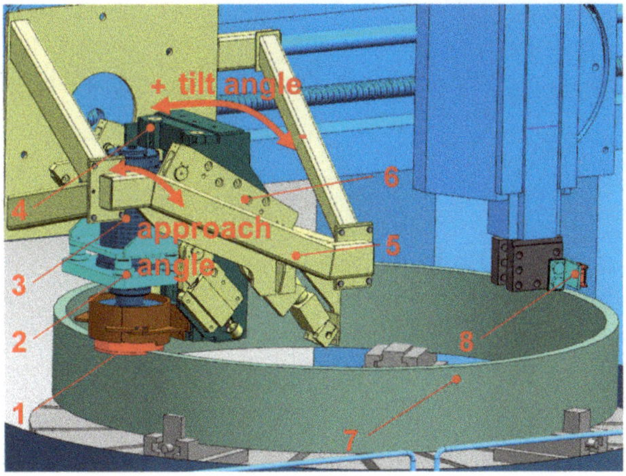

Figure 1. Experimental set-up: 1 grinding wheel, 2 force measurement system, 3 spindle motor, 4 vertical pneumatic cylinder, 5 frame, 6 milled slide, 7 workpiece, 8 laser sensor.

Tilting the whole grinding spindle around the y-axis, as shown in Figure 2, adjusts the tilt angle while the approach angle of the grinding unit is connected to rotation around the x-axis.

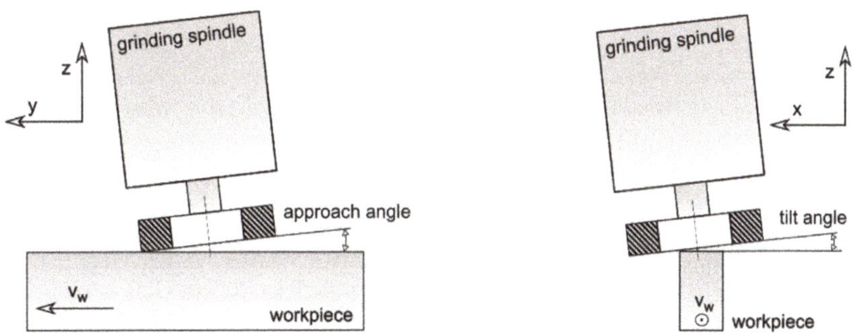

(**a**) (**b**)

Figure 2. Schematic representation of the approach angle (**a**) and tilt angle (**b**).

Approach angle and tilt angle can be adjusted only in correlation to each other. A typical kinematic interaction between the grinding wheel and the workpiece is a single-sided contact. For the tilt angle $-1.5°$, the corresponding approach angle reaches zero leading to a dual-sided contact of the grinding wheel. This fundamental kinematic difference depending on the tilt angle is shown in Figure 3. Different surface patterns on the workpiece can be observed for both kinematic situations as shown later. Additionally, the self-sharpening behaviour for the dual-sided contact is significantly lowered due to the pressure being split on the two areas. Material removal rate and grinding wheel wear change between both kinematic situations.

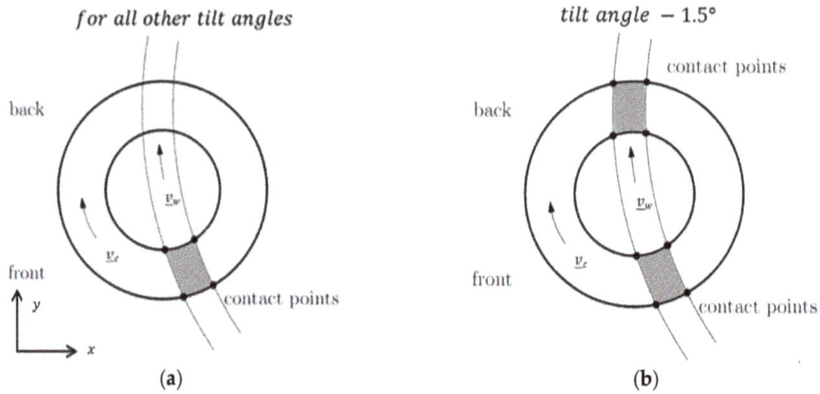

Figure 3. One-sided contact (**a**) and double-sided contact; (**b**) of the grinding wheel after initial self-sharpening.

3. Modelling

3.1. Stochastic Process Model

Based on the kinematic simulation presented in [17–19] a kinematic model with implemented wear model is developed for resin bonded grinding wheels. The general scheme of the stochastic process simulation with its main steps is shown in Figure 4.

The main input of the simulation is the grinding wheel geometry. Its generation is presented and follows in six steps: the generation of the basic grain form, the modification of the grain shape, the grain density definition, the extraction of the cutting profile, the positioning of the grain at the grinding wheel surface, and the height distribution of the grain. The grinding wheel macro geometry is defined by a circle formula. For abrasive grains modelled, five basic forms: cuboid, triangular prism, two triangular pyramid, and dodecahedron are defined. These are re-shaper by edge and corner fracture. This shape modification leads to a large amount of different abrasive grains corresponding with the real grinding wheel, where crushed corundum abrasive grains are used. The abrasive grain density is estimated experimentally leading and used as input for the grinding wheel generation. For the abrasive grain height distribution, a triangular distribution delivers the most accurate results, compared with uniform and normal distribution, and it is used in the grinding wheel topography modelling. The resulting modelled grinding wheel topography is then compared with the topography of the grinding wheel used experimentally by Abbott-Firestone curve. A good agreement can be found as shown in Figure 5. The deviation towards the higher profile ratio is due to the modelled grinding wheel, where only one layer of the abrasive grains is taken into account and the modelling concentrates on the cutting edge space that comes into contact with the material. The deviation in the valleys of the grinding wheel due to the approach of monolayer does not play any role for the process. The workpiece geometry is meshed with nodes in radial and tangential direction. The nodes can be chosen arbitrarily. In the meshed workpiece geometry, the resulting penetration between workpiece and grinding wheel grains is saved as a surface roughness. The result quality is directly dependent on node quantity.

Process parameters that are used in the experiment and listed in the previous section are used also in the simulation. The kinematic description used in the simulation is taken from [17].

After the definition of all the input parameters from Figure 4, including the grinding wheel, the kinematic simulation can be performed. Abrasive grains, which cannot be in contact with the workpiece, are deleted and only possible kinematic active grains are used for further calculation. This allows, particularly for small cutting depths, a huge reduction of computational time.

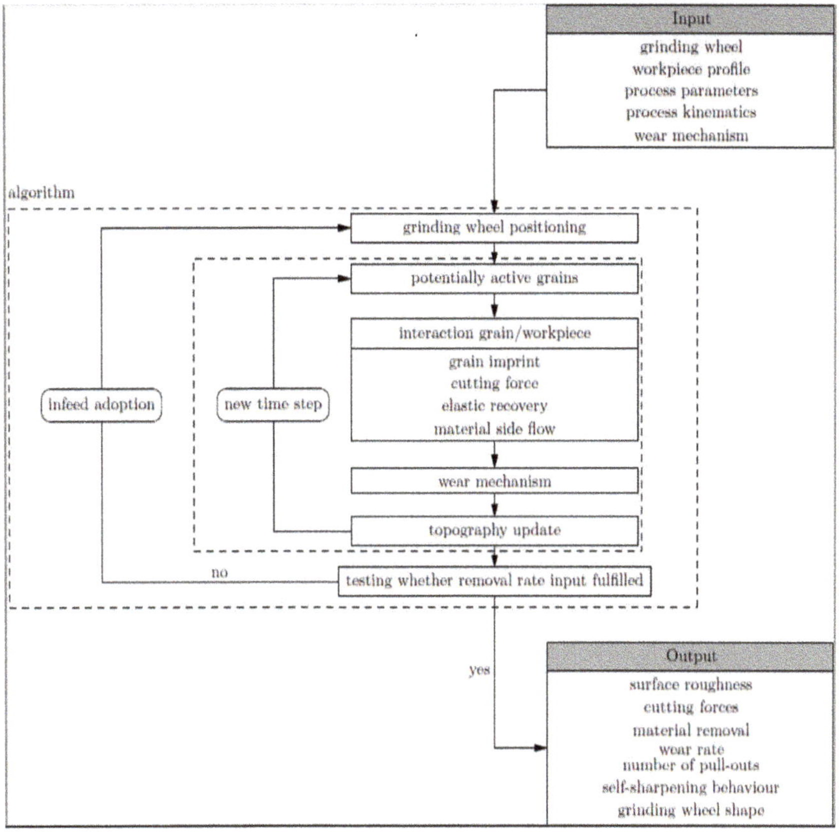

Figure 4. Structure of the kinematic process simulation.

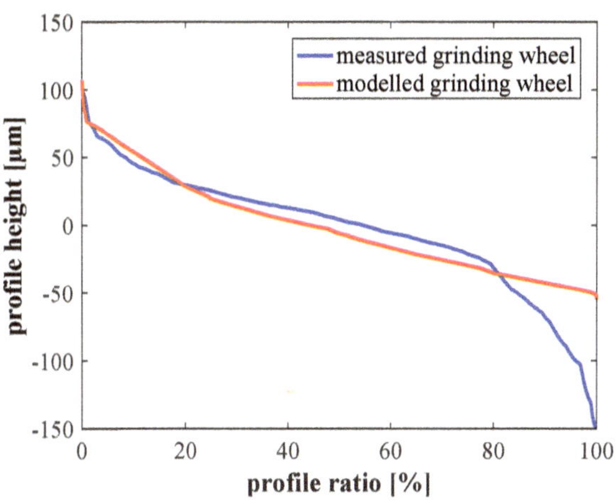

Figure 5. Comparison of grinding wheel topographies for measured and modelled grinding wheel by Abbott-Firestone curve.

In the simulation an ideal grinding wheel shape is loaded. According to the defined approach angle, the geometry needs to be adapted. The dressed grinding wheel geometry is determined by the contact points of this idealised grinding wheel with the workpiece profile. All of the cutting points between the grinding wheel zero plane at outer and inner wheel diameter and between the workpiece profile need to be found. The approach angle can be gained by contact point height difference at the inner and outer grinding wheel diameter. In Figure 6, possible contact points are shown for grinding process where the dressing angle α is defined.

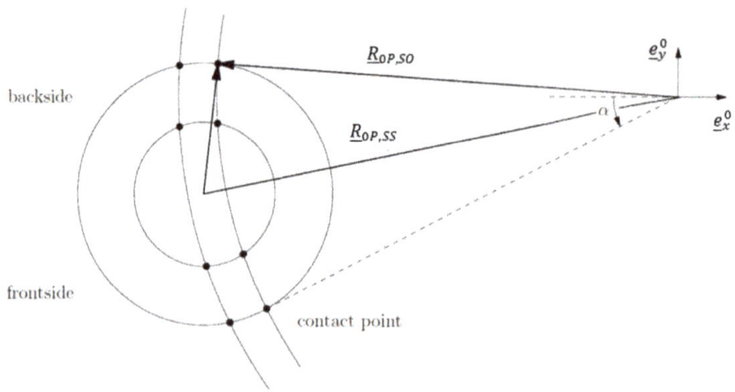

Figure 6. Kinematic situation for double-sided contact between the grinding wheel and workpiece with the connecting vectors.

The positioning vector of the contact point can be calculated in the workpiece and grinding wheel coordinate system. Transforming both positioning vectors in the global coordinate system and equalising them, an equation for each contact point can be defined. A cutting point on the grinding wheel surface follows according to the equation:

$$\underline{R}_{OP,SS} = \underline{R}_{OB} + \underline{\underline{A}}_{0B} \cdot \underline{R}_{BC} + \underline{\underline{A}}_{0B} \cdot \underline{\underline{A}}_{BC} \cdot (\underline{R}_{CD} + \underline{R}_{DP}) \tag{1}$$

where the SS subscript indicates that point P is the cutting point from a grinding wheel point of view. $\underline{R}_{OP,SS}$ is the connecting vector between an arbitrary point P at the grinding wheel surface and the origin of the coordinate system indicated in Figure 6. $\underline{R}_{OB}$ is the connecting vector from the origin to the grinding wheel centre, $\underline{R}_{BC}$ is the connecting vector from grinding wheel centre to the tilting point, $\underline{R}_{CD}$ is the connecting vector from the tilting point to the already tilted centre of the grinding wheel, and $\underline{R}_{DP}$ is the connecting vector from the tilted grinding wheel centre to the arbitrary point on the grinding wheel surface P. The nomenclature of the vector is as follows: subscript describes the connection between two points in a coordinate system. $\underline{\underline{A}}_{0B}$ and $\underline{\underline{A}}_{BC}$ are then transformation matrices defined as:

$$\underline{\underline{A}}_{0B} = \begin{pmatrix} \cos(\beta-\beta_0) & -\sin(\beta-\beta_0) & 0 \\ \sin(\beta-\beta_0) & \cos(\beta-\beta_0) & 0 \\ 0 & 0 & 1 \end{pmatrix} \tag{2}$$

and

$$\underline{\underline{A}}_{BC} = \begin{pmatrix} \cos\Omega_K & \sin\Omega_A \sin\Omega_K & \cos\Omega_A \sin\Omega_K \\ 0 & \cos\Omega_A & -\sin\Omega_A \\ -\sin\Omega_K & \sin\Omega_A \cos\Omega_K & \cos\Omega_A \cos\Omega_K \end{pmatrix} \tag{3}$$

where Ω_A is the approach angle and Ω_K is the tilt angle. The transformation matrix $\underline{\underline{A}}_{0B}$ describes the rotation of the local reference system B to the global coordinate system and $\underline{\underline{A}}_{BC}$ is responsible for the adjustment of both, the tilt and approach angle.

The interaction between a pre-ground workpiece and a single grain is shown in Figure 7. From the irregular cross section in cutting speed direction multiple overlaps with the cutting surface of the grain might arise.

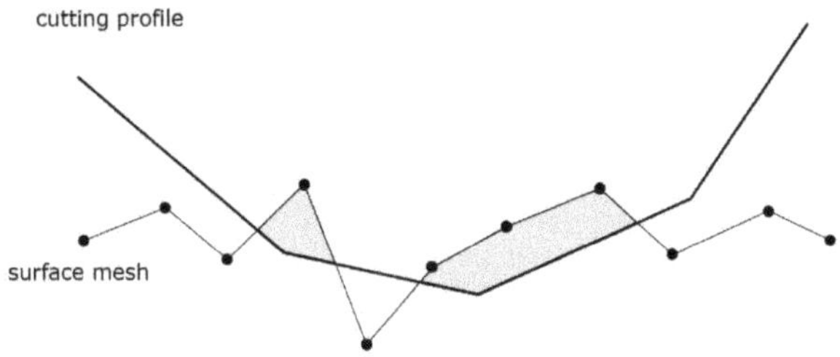

Figure 7. Intersection between one abrasive grain cutting profile and already ground workpiece surface cross section seen in the direction of the cutting speed.

For every polygon, the cutting area can be calculated and summed up to a total cutting area for each abrasive grain. This is used then as input for the force calculation. To calculate the force for the whole grinding wheel, information about the force acting points for each grain is necessary. It is assumed that this acting point is in the centre of gravity of the respective cutting area. The radial position r and height position h of the centre of gravity are calculated with respect to the grinding wheel referential plane as:

$$r_t = \frac{1}{6 \cdot A_t} \cdot \sum_{i=1}^{i-1} [r(i) + r(i+1)] \cdot [(h(i) - h(i+1)) \cdot (r(i) \cdot r(i-1))] \tag{4}$$

$$h_t = \frac{1}{6 \cdot A_t} \cdot \sum_{i=1}^{i-1} [h(i) + h(i+1)] \cdot [(h(i) - h(i+1)) \cdot (r(i) \cdot r(i-1))] \tag{5}$$

A_t is the partial polygon area from the intersection between the abrasive grain and the workpiece. The centre of gravity can be then calculated from the partial polygon areas as:

$$r = \frac{1}{A} \cdot \sum_{j=1}^{i} r_t \cdot A_t \tag{6}$$

$$h = \frac{1}{A} \cdot \sum_{j=1}^{i} h_t \cdot A_t \tag{7}$$

The cutting force can be calculated according to an equation similar to the Kienzle equation as:

$$F_c = k_{c1,1} \cdot A_{cut}(t) \tag{8}$$

where $k_{c1,1}$ is the experimentally evaluated specific force and A represents the abrasive grain area that is engaged with the workpiece and is orthogonal to the cutting direction. The force magnitude delivered by the adapted Kienzle equation needs to be complemented by its direction. From the 3D abrasive grain geometry, as described in [17], the orthogonal cross-sectional cutting line is derived. In addition to this, the projection of all other grain faces in contact with the workpiece on the orthogonal cutting line is calculated. Each of these projected faces has its normal vector associated with it. This allows the 3D abrasive grain geometry simplification without losing information. Combining these projected normal vectors for one abrasive grain gives a resulting force direction on one abrasive grain:

$$\underline{F_i} = \sum_{i=1}^{i_{\max}} \frac{v_i}{v_t} \cdot \underline{n_i} \tag{9}$$

where $i_{\max}$ is the total number of projected normal faces of one cutting grain, v_i is the volume extracted by the single face i, v_t is the total volume extracted by the grain in the given time step, and $\underline{n_i}$ is the normal vector associated with the face i. This resulting force can be split into the cutting force F_c, showing in the direction of the cutting speed and the normal force F_N being perpendicular to the cutting force.

The resulting cutting force for the whole grinding wheel is then the sum of all the partial forces acting:

$$\underline{F_c} = \sum_{i=0}^{n} \underline{F_{ci}} \tag{10}$$

The normal forces are then calculated as:

$$F_N = \frac{F_c}{\mu} \tag{11}$$

where the friction coefficient $\mu = 0.3$ is estimated from the experiment.

As a power controlled spindle is used in the experiment, the proper depth of cut needs to be calculated. As explained above, the nominal current of the grinding spindle electric engine is set and compared with the actually measured current. The measured current is a result of the contact condition between the grinding wheel and the workpiece, and thus of the axial force on the grinding wheel exerted by a pneumatic cylinder. This force needs to be counterbalanced by the sum of the penetration forces of the individual active grains. With increasing force, the amount of active abrasive grains and their penetration depth increases and the current rises. For each height of the grinding wheel, the penetration depth of every single stochastically distributed, oriented, and shaped active grain can be calculated, and thus the forces can be derived from the single grain force model. Iteration is then performed until the sum of the individual grain forces equals the externally applied axial force, pneumatic, and dead weight force, on the spindle. The resulting experimentally measured vertical force is used in the simulation as the nominal value. An estimate of the number of active abrasive grains is known, and thus the force can be distributed among them. The initial grinding wheel position is defined as one where the highest abrasive grain touches the workpiece surface. The simulation is performed with a stepwise increase of the depth of the cut by the amount of one percent of an average abrasive grain diameter, as schematically shown in Figure 8. If the sum of all vertical forces of the individual abrasive grains in contact with the workpiece is equal to the nominal, experimental, force value, then the simulation continues at this depth of cut. As the propagated wear of the abrasive grain can lead to an unbalance between the nominal and the real vertical force, the depth of cut needs to be adjusted iteratively. In case an abrasive grain is completely worn and is out of contact, the resulting forces are lowered and the depth of cut needs to be increased again. The worn abrasive grain is in contact again and, additionally, new abrasive grains are potentially cutting. The stepwise increase or decrease of the depth of cut is done until the balance between the nominal and simulated force is found again. When the equilibrium between the simulated and experimental forces is reached, the calculation of the workpiece surface roughness can start.

The workpiece material side flow and the workpiece spring back are modelled to determine the ground profile. Based on Li [20], the side burr is modelled as a triangular ridge accompanying the scratch. It is described by its base b and height h. Li [20] claims that the side flow is not significantly affected by the abrasive grain rake angle and that the triangle base b is three times its height h which gives for h:

$$h = 1.27 \cdot {A_{cut}}^{0.29} \cdot v_g^{-0.87} \quad (12)$$

where v_g is the grain velocity in m/s and A_{cut} the area of the cutting profile being engaged with the workpiece in m^2.

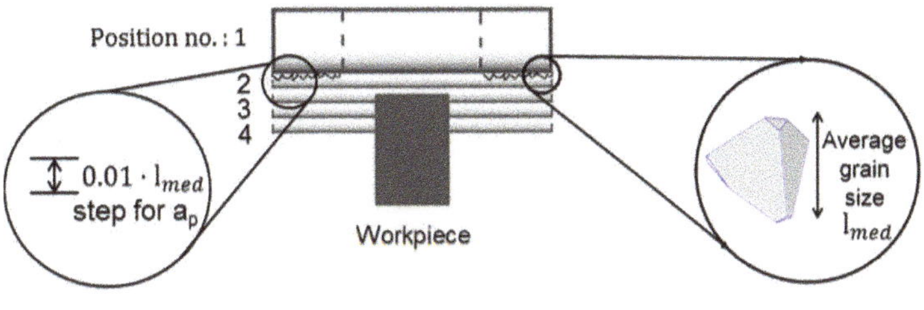

Figure 8. A stepwise adjustment of the depth of the cut in the simulation.

The material spring back model is based on a publication by Shaw and DeSalvo [21] about hardness measurements. They claim that the material actively reacting on the imposed pressure on the surface is maximum ten times the penetration of the indenter. The elastic material displacement can be calculated as:

$$\Delta l = \frac{F \cdot l}{E \cdot A} \quad (13)$$

where Δl is the elastic material displacement, F is the normal force acting on the workpiece surface, E is the elastic modulus, l is the material depth responding to the abrasive grain penetration, and A is the workpiece area responding to the applied pressure. The force from experiment is in the range from 1200 to 1400 N. This force is then divided between all of the active abrasive grains and an average force is attributed to a single active grain. The length l is attributed to the workpiece material amount, which answers to the pressure applied.

3.2. Wear Modelling

The main wear mechanism identified as responsible for the geometry change of the grinding wheel is a fracture of the abrasives. This is a result of an initial abrasive grain flattening causing force increase until fracture.

Grain fracture is observed experimentally, after reaching steady state, and is mainly in a vertical direction—the direction of the normal force. The vertical component comes from the high normal pressure due to the spindle weight and the power control. The horizontal component is resulting from the cutting process and thus the spindle torque.

The main parameters influencing the grinding wheel wear are the cutting force, the cutting speed, and the contact temperature. With increasing active time of the abrasive grain, the wear is propagating until a critical load is reached. This behaviour can be expressed as:

$$\frac{dW}{dt} = K \cdot F_c(t) \cdot v_c \cdot e^{-\frac{B}{T_{grind}}} \quad (14)$$

where dW/dt is the wear rate in mm/s, K is the wear factor, B is a constant coming from the Arrhenius equation, where activating energy and universal gas constant are considered, F_c is the cutting force, v_c is the cutting speed for the abrasive grain, and T_{grind} is the contact temperature between the abrasive grain and the workpiece. Evaluation of the wear factor K is based on an experimental macroscopic wear evaluation.

The cutting area A_{cut} in Equation (8) changes with the wear progress $W(t)$ and can be calculated as:

$$A_{cut}(h(t)) = [(h_0 - W(t)] \cdot b \tag{15}$$

where h_0 is the initial abrasive grain height and b is the abrasive grain width assumed to be constant during the whole cutting process as shown in Figure 9.

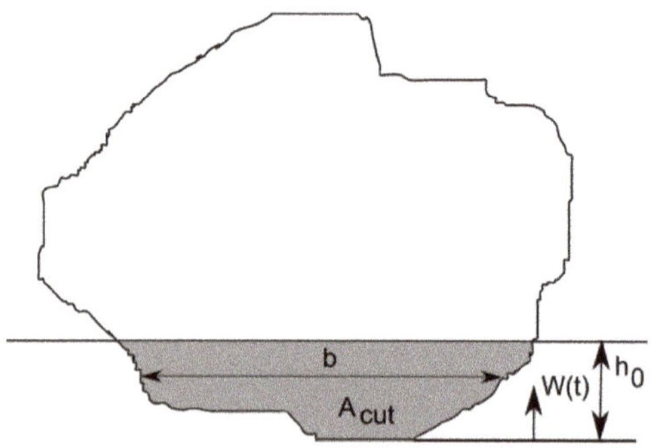

Figure 9. Abrasive grain geometry with parameters used for wear calculation.

The cutting force also depends on the cutting edge condition, which by wear becomes duller. This leads to cutting force increase and can be considered by the introduction of a cutting edge dulling factor M. The force can be then calculated as:

$$F_c = k_{c1,1} \cdot \left(1 + M \cdot \frac{W(t)}{h_0}\right) \cdot A_{cut}(h(t)) \tag{16}$$

With the cutting area being calculated according to Equation (15), the cutting force can be calculated as:

$$F_c = k_{c1,1} \cdot b \cdot \left(h_0 - W(t) + M \cdot W(t) - M \cdot \frac{W^2(t)}{h_0}\right) \tag{17}$$

Substituting Equation (17) into Equation (14) leads to:

$$\frac{dW}{dt} = K \cdot v_c \cdot e^{\left(-\frac{B}{T_{grind}}\right)} \cdot k_{c1,1} \cdot b \cdot \left(h_0 - W(t) + M \cdot W(t) - M \cdot \frac{W^2(t)}{h_0}\right) \tag{18}$$

The whole term in front of the bracket is a constant in time and is renamed by H leading to:

$$\frac{dW}{dt} = H \cdot \left(h_0 - W(t) + M \cdot W(t) - M \cdot \frac{W^2(t)}{h_0}\right) \tag{19}$$

This differential equation only for the wear and can be analytically solved to:

$$W(t) = h_0 \cdot \frac{e^{c_1 h_0 M + c_1 h_0 + HMt + Ht} + 1}{e^{c_1 h_0 M + c_1 h_0 + HMt + Ht} - M} \tag{20}$$

where c_1 is a constant. As $W(0) = 0$ leads to:

$$c_1 = \frac{\ln(-1/h_0)}{h_0 \cdot (M+1)} \tag{21}$$

and thus to the solution:

$$W(t) = h_0 \cdot \frac{e^{Ht(M+1)} - 1}{e^{Ht(M+1)} + M} \tag{22}$$

The contact temperature between the abrasive grain and the workpiece defining H, given by comparison between Equations (18) and (19) is calculated according to [22] as:

$$T_{grind} = 3.1 \cdot \frac{2 \cdot R_w \cdot q_f \cdot \alpha}{\pi \cdot k_T \cdot v_w} \cdot \left(\frac{v_w \cdot l}{2 \cdot \alpha}\right)^{0.53} \tag{23}$$

where R_w is the energy partition coefficient, q_f is the heat flux, α is the workpiece thermal diffusivity, k_T is the thermal conductivity of the workpiece, v_w is the workpiece feed rate, and l is the half length of the band source.

A criterion needs to be established for the grain failure. Stresses acting on the abrasive grain are estimated and compared to the critical stress value. The stress in the cutting direction can be estimated as:

$$\sigma_{cut} = \frac{F_c}{A_{cut}(h(t))} \tag{24}$$

The stress in the normal direction can be calculated as:

$$\sigma_N = \frac{F_N}{A_\perp} \tag{25}$$

where F_N is the normal force and $A_\perp$ is the cross section of the abrasive grain projected in normal direction. The normal force can be calculated from Equation (11).

Additionally, a shear stress can be calculated as:

$$\tau_{yz} = \frac{F_c}{b^2} \tag{26}$$

where F_c is the cutting force and b the width of abrasive grain.

For the calculation of the equivalent stress, the Rankine criterion is used:

$$\sigma_{result} < \sigma_{critical} = 300 \text{ MPa} \tag{27}$$

Plotting this resulting stress σ_{result} qualitatively for different wear factors K and with M = 1 as simplifying assumption, different progressions can be observed as shown in Figure 10.

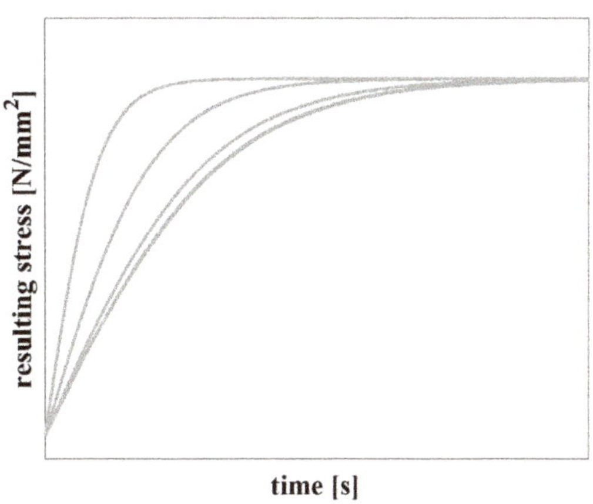

Figure 10. Qualitative representation of the resulting stress for a single abrasive grain for different wear factors K.

From Figure 10 and together with the knowledge of the experimentally evaluated average time for an abrasive grain failure and the calculated resulting stress, the wear factor can be evaluated. Wear factor $K = 508 \times 10^{-4}$ 1/N is estimated as shown in Figure 11.

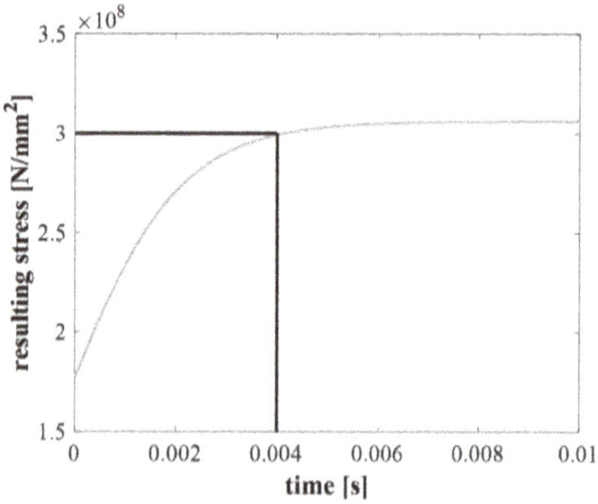

Figure 11. Resulting overall stress with the evaluated wear factor *K*.

With the wear factor *K*, the propagation of wear with time can be calculated according to Equation (22).

4. Results and Discussion

In order to compare and validate the simulation with the experiment, the workpiece surface roughness measurements are done. An indirect measuring method is carried out for all of the roughness measurements. An imprint mass used for dental application is uniformly distributed on the workpiece surface by dispense pistol, as shown in Figure 12. The overall imprint length allows for the maximum measurement length of 80 mm. This imprinting method is time effective and allows for the conservation of the current surface ground for later evaluation without losing the information. The accuracy of this imprinting measuring method is proven in [23].

Figure 12. Workpiece surface imprint used for indirect roughness measurements.

All of the roughness measurements on the imprints are performed with a tactile roughness measurement device Talysurf from Taylor Hobson, using diamond tip with 2 μm tip radius and cut-off filter 2.5 mm. The measuring direction is along the feed rate direction, perpendicular to the grinding grooves.

For all of the grinding experiments, the following process parameters were used: cutting speed v_c = 47 m/s, feed rate v_w = 120 m/min, and grinding time 6 s. The grinding wheel was in steady state. Pre-dressing, using the self-sharpening effect of the grinding wheel, was performed prior to every experiment.

Figure 13 compares experimental and simulated results between the workpiece surface topography for tilt angle 0.0°. Good agreement between the experimental and simulated roughness values can be achieved. The roughness values *Ra* and the mean roughness value *Rz* are summarised in Table 1. As validation of the model, the comparison of *Ra* and *Rz* might be used. For the purpose of this research it is sufficient to verify the prediction of surface properties by *Ra* and *Rz* roughness values. The workpiece surface anisotropy is in feed rate direction and in the cutting speed direction. As the workpiece surface roughness, by railway grinding, in the feed rate direction is of interest, *Ra* and *Rz* are sufficient.

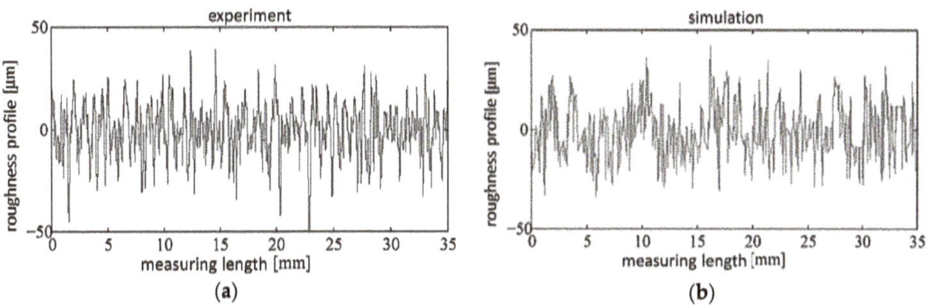

Figure 13. Comparison of surface topography for the experiment (**a**) and the simulation (**b**) using the tilt angle 0.0° and the approach angle 0.12°.

Table 1. Average roughness *Ra* and average roughness depth *Rz* for experiment and simulation.

Tilt Angle	Average Roughness *Ra* (µm)		Average Roughness Depth *Rz* (µm)	
	Experiment	Simulation	Experiment	Simulation
2.5°	9.8	9.9	61.8	59.4
0.0°	9.5	9.4	61.7	55.7
−1.5°	5.3	6.6	35.7	39.9

It can be seen, that for the tilt angle −1.5° in Figure 14, where the grinding wheel is in dual-sided contact with the workpiece, the second part of the grinding wheel ruptures the roughness peaks. The workpiece surface and the corresponding roughness values are lower as for the tilt angle 0.0°. When considering the workpiece surface topology, a crisscross-like pattern can be observed for the tilt angle −1.5°.

The mean roughness *Ra* reaches *Ra* = 6.6 µm for the simulation and *Ra* = 5.3 µm for the experiment. The mean roughness depth *Rz* reaches *Rz* = 39.9 µm for the simulation and *Rz* = 35.7 µm for the experiment.

When comparing the average *Ra* and *Rz* roughness values for tilt angles −1.5°, 0.0°, and 2.5° in Table 1, the simulation results at a tilt angle of 2.5° differ by maximum 12.2% for the *Ra* values and by 14.4% for the *Rz* values. For the tilt angle 0.0°, the deviation by the *Ra* values is maximum 5.2% and by the *Rz* values is 11%. For the double-sided grinding wheel contact, where the tilt angle is −1.5°, the deviation for *Ra* is 24% and for *Rz* is 12%. This is caused by the associated approach angle not being equal to zero. The self-sharpening effect of the resin bonded corundum grinding wheel causes grinding wheel adaption on the workpiece geometry. Despite the relatively large difference in the process kinematics for these both tilt angles, one size of the grinding wheel cutting surface is always planar on the ground surface. This condition causes similar roughness values for all different tilt approaches, except the one where the approach angle is zero (only for tilt angle −1.5°).

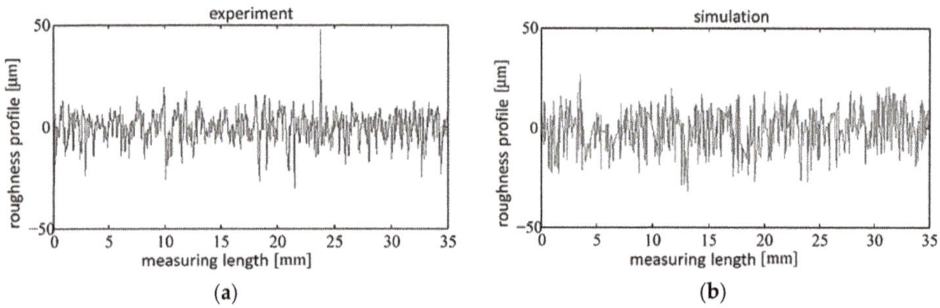

Figure 14. Comparison of surface topography for the experiment (**a**) and the simulation (**b**) using the tilt angle −1.5° and the approach angle 0.01°.

5. Conclusions

A kinematic process model with the focus on the wear modelling was presented. Time dependent wear behaviour was presented. High feed rates were used for all of the experiments. The dependence of surface roughness on the kinematic situation could be shown. With a dressed grinding wheel, any adjusted tilt angle between ±15° shows the same workpiece roughness. This is because of the self-sharpening of the resin bonded corundum grinding wheel and the adaption to the workpiece geometry. The self-sharpening effect of the corundum grinding wheel shows that only the approach angle 0.0° can improve the surface roughness, where a dual-sided grinding wheel contact is used.

An analytical approach for the wear modelling has been chosen based on previous studies. The wear rate has been modelled as a function of the cutting force, the cutting speed, and the temperature for every abrasive grain. The resulting exponential behaviour of the wear propagation was used for subsequently determining the resulting stress on the abrasive grain, which is compared to some critical stress level to determine the grain failure. With this model, the time delayed failure of a grain could be described. Three wear parameters K, M, and B were introduced based on experimental investigations. The average time for a grain failure was calculated and the wear factor $K = 508 \times 10^4$ 1/N was estimated.

The simulation results show good agreement with the measured workpiece surface roughness. When comparing the roughness values Ra and Rz, the deviation is by maximum 14.4% for tilt angle 2.5°, 11% for 0.0° and 24% for −1.5°.

Acknowledgments: The authors would like to thank the Commission for Technology and Innovation (CTI), the Federal Office for the Environment (FOEN) and the Federal Office of Transport (FOT) for the fruitful cooperation which made this work possible.

Author Contributions: Michal Kuffa conceived, designed and performed the experiments, analyzed the data and wrote the manuscript. Fredy Kuster contributed to the analysis and the interpretation of results. Konrad Wegener contributed to the modelling part, discussion and reviewed and approved the manuscript.

Conflicts of Interest: The authors declare no conflict of interest.

References

1. Bhaduri, D.; Chattopadhyay, A.K. Influence of Grinding Parameters and Substrate Bias Voltage in Dry Surface Grinding with TiN-Coated Single Layer Galvanic cBN Wheel. *Mater. Manuf. Process.* **2011**, *26*, 982–990. [CrossRef]

2. Kopac, J.; Krajnik, P. High-performance grinding—A review. *J. Mater. Process. Technol.* **2006**, *175*, 278–284. [CrossRef]
3. Schriefer, H. *Kontinuierliches Wälzschleifen von Verzahnungen*; Reishauer: Wallisellen, Switzerland, 2008.
4. Aurich, J.C.; Herzenstiel, P.; Sudermann, H.; Magg, T. High-performance dry grinding using a grinding wheel with a defined grain pattern. *CIRP Ann.* **2008**, *57*, 357–362. [CrossRef]
5. Jackson, M.J.; Hitchiner, M.P. *High Performance Grinding and Advanced Cutting Tools*; Springer: New York, NY, USA, 2013.
6. Rabiey, M.; Jochum, N.; Kuster, F. High performance grinding of zirconium oxide (ZrO_2) using hybrid bond diamond tools. *CIRP Ann.* **2013**, *62*, 343–346. [CrossRef]
7. Tawakoli, T.; Tavakkoli, S.J. High-Efficiency Deep Grinding (HEDG) of Inconel and Other Materials. In *Superabrasives*; General Electric: Worthington, OH, USA, 1991; Volume 91, pp. D67–D82.
8. Chen, Z.Z.; Xu, J.H.; Ding, W.F.; Cheng, Z.; Fu, Y.C. High Speed Grinding of Nickel-Based Superalloy with Single Diamond Grit. *Adv. Mater. Res.* **2011**, *325*, 140–146. [CrossRef]
9. Rowe, W.B.; Jin, T. Temperatures in high efficiency deep grinding (HEDG). *CIRP Ann.* **2001**, *50*, 205–208. [CrossRef]
10. Hoffmeister, H.W. Hohe Zerspanungsleistungen Durch Schleifen Mit CD (Continuos Dressing) Sichere, Werkstoffangepasste und Wirtschaftliche Prozessführung. Master's Thesis, Technischen Universität Carolo-Wilhelmina zu Braunschweig, Braunschweig, Germany, 1995.
11. Burkhard, G.; Rehsteiner, F.; Schumacher, B. High efficiency abrasive tool for honing. *CIRP Ann.* **2002**, *51*, 271–274. [CrossRef]
12. Linke, B.S. Review on Grinding Tool Wear With Regard to Sustainability. *J. Manuf. Sci. Eng.* **2015**, *137*, 060801. [CrossRef]
13. Jiang, J.L.; Ge, P.Q.; Bi, W.B.; Zhang, L.; Wang, D.X.; Zhang, Y. 2D/3D ground surface topography modeling considering dressing and wear effects in grinding process. *Int. J. Mach. Tools Manuf.* **2013**, *74*, 29–40. [CrossRef]
14. Liu, Y.; Warkentin, A.; Bauer, R.; Gong, Y. Investigation of different grain shapes and dressing to predict surface roughness in grinding using kinematic simulations. *Precis. Eng.* **2013**, *37*, 758–764. [CrossRef]
15. Kacalak, W.; Kasprzyk, M.; Krzyżyński, T. On Modelling of Stochastic Processes of Abrasive Wear and Durability of Grinding Wheel. *PAMM* **2003**, *2*, 278–279. [CrossRef]
16. Meng, H.C.; Ludema, K.C. Wear models and predictive equations: Their form and content. *Wear* **1995**, *181–183*, 443–457. [CrossRef]
17. Kuffa, M.; Züger, S.; Kuster, F.; Wegener, K. A Kinematic Process Model and Investigation of Surface Roughness for High Efficiency Dry Grinding. *Procedia CIRP* **2016**, *46*, 636–639. [CrossRef]
18. Vargas, G.E. Analyse und Simulation des Prozesses Honräumen von Gehärteten Innenprofilen mit Diamantwerkzeugen. Ph.D. Thesis, Eidgenössische Technische Hochschule ETH Zürich, Zürich, Switzerland, 2010.
19. Pinto Wagner, F. An Experimental and Numerical Approach to Investigate the Machining Performance of Engineered Grinding Tools. Ph.D. Thesis, Eidgenössische Technische Hochschule ETH Zürich, Zürich, Switzerland, 2008.
20. Li, X. Modeling and Simulation of Grinding Processes Based on a Virtual Wheel Model and Microscopic Interaction Analysis. Ph.D. Thesis, Manufacturing Engineering, Worcester Polytechnic Institute, Worcester, MA, USA, 2010.
21. Shaw, M.C.; DeSalvo, G.J. On the Plastic Flow Beneath a Blunt Axisymmetric Indenter. *J. Eng. Ind.* **1970**, *92*, 480–492. [CrossRef]

22. Kuffa, M.; Kuster, F.; Wegener, K. Comparison of lubrication conditions for grinding of mild steel with electroplated cBN wheel. *CIRP J. Manuf. Sci. Technol.* **2016**, *18*, 53–59. [CrossRef]
23. Henerichs, M.; Egeter, M.; Liebrich, T.; Voss, R.; Wegener, K. Evaluation of the IWF-Wunder Reproduction Method for Generating Positive Replica. *Int. J. Autom. Technol.* **2014**, *8*, 49–56. [CrossRef]

inventions MDPI

Article

Design Process Control for Improved Surface Finish of Metal Additive Manufactured Parts of Complex Build Geometry

Mikdam Jamal [1] and Michael N. Morgan [2,*]

1 The Manufacturing Technology Centre Ltd., Ansty Park, Coventry CV7 9JU, UK; Mikdam.Jamal@the-mtc.org
2 Advanced Manufacturing Technology Research Laboratory (AMTREL), Faculty of Engineering and Technology (FET), Liverpool John Moores University, Liverpool L3 3AF, UK
* Correspondence: m.n.morgan@ljmu.ac.uk; Tel.: +44-(0)151-2312590

Received: 11 August 2017; Accepted: 6 December 2017; Published: 13 December 2017

Abstract: Metal additive manufacturing (AM) is increasingly used to create complex 3D components at near net shape. However, the surface finish (SF) of the metal AM part is uneven, with surface roughness being variable over the facets of the design. Standard post-processing methods such as grinding and linishing often meet with major challenges in finishing parts of complex shape. This paper reports on research that demonstrated that mass finishing (MF) processes are able to deliver high-quality surface finishes (Ra and Sa) on AM-generated parts of a relatively complex geometry (both internal features and external facets) under select conditions. Four processes were studied in this work: stream finishing, high-energy (HE) centrifuge, drag finishing and disc finishing. Optimisation of the drag finishing process was then studied using a structured design of experiments (DOE). The effects of a range of finishing parameters were evaluated and optimal parameters and conditions were determined. The study established that the proposed method can be successfully applied in drag finishing to optimise the surface roughness in an industrial application and that it is an economical way of obtaining the maximum amount of information in a short period of time with a small number of tests. The study has also provided an important step in helping understand the requirements of MF to deliver AM-generated parts to a target quality finish and cycle time.

Keywords: metal additive manufacturing; mass finishing; process optimization

1. Introduction

Presently, the surface finish of non-machined complex near net shape metal additive manufacturing (AM) components is rough and uneven, having a relationship with build orientation, machine build parameters, material and powder size. This is an increasingly challenging problem as AM components are frequently of high complexity, rendering them even more difficult to finish, which deleteriously impacts on post-processing time and labour, driving costs upwards. The relatively poor surface finish is recognised as a barrier to wider adoption by industry and prevents fuller exploitation of this disruptive technology. Mass finishing (MF) processes are often considered for finishing AM parts, but it is challenging to know the appropriate abrasive media grade, size, shape, process parameters, fixturing system and controls to deliver the desired ouptut.

A review of the literature shows that only limited scientific research has been published with respect to mass finishing, more specifically associated with AM, and few mathematical or process models exist. Most notably, Gillespie [1] is recognised as one of the earliest and major contributors in this field, with his book being used as a guide to research. Further important publications [2,3] are a reference for the scientific community and professionals alike. A more recent key paper [4] on topographical development and areal characterization provides guidance on the optimal parameters and sampling method to characterise this surface type for a given application. A number of trade articles and empirical studies have also been published [5–8]. These describe, in general terms, how surface finish and material removal are controlled, but these have only limited usefulness as they provide little insight into the comparative performance of different tool materials; they fail to explain fully and scientifically the relationships between key process parameters.

It is reasoned that, with the benefit of increased data, the relationship between finishing parameters and surface finish could be predicted more accurately, thus leading to more robust design tools that could serve to improve quality in the AM process and the final component. The introduction of this design aims to advance both AM and surface engineering technologies, with the ability to enhance AM component surface finish quality through post-processing methodologies to suit the AM parts. Post-processing costs and AM manufacturing costs will thereby be reduced, increasing the competiveness of the AM manufacturer and the surface finishing company and leading to increased profitability. Achieving surface finish specification will allow other industries to increase their uptake of metal AM, utilise increased part complexity in their processes, and take advantage of open surface coating opportunities for these parts.

The aims of the study were to obtain a quantitative understanding of the capability and efficacy of MF processes in delivering an AM part with a target surface finish over a range of system input conditions and to create a dataset and inferred rules based on process informatics. This would serve as a generic platform to aid the user with selecting optimal production parameters for the generation of parts with a specified output surface finish and/or production cost.

2. Mass Finishing Process Review

There are a number of different MF processes in common use in industry. Among these are barrel, centrifugal and vibratory. Vibratory systems have become the dominant technique due to the advantages inherent to the method in terms of ease of use and material handling. There is no fixturing in these systems, so the parts can flow freely with the media. Such systems fall into two broad categories in terms of the equipment being used: round bowl and rectangular tub designs [9]. These systems are typically modest in size and used for the deburring and finish processing of smaller components. The capability to process larger batch lots of various sized parts is important to the competitiveness of such processes. The mass finishing processes are described in the following sections.

2.1. Mass Finishing Processes

2.1.1. Centrifugal Disc

Centrifugal disc machines were designed to combine the three-dimensional action advantage of a vibratory bowl with the rapid cycle time of the centrifugal barrel. Figure 1 shows an image of the principle of a centrifugal disc machine. The centrifugal force and the rotating action of the disc sends the mass outwards towards the chamber wall. The mass then slows down and returns to the centre to complete the cycle [10]. Centrifugal disc machines are also used for their fast processing and cycle time (typically 10 to 30 min), their capability for batch production and their control in process [1]. However, due to the small capacity of the chamber, the workpiece should not exceed 30 cm in length.

2.1.2. High-Energy Centrifugal Barrel

Centrifugal barrel machines often provide finishing results that cannot be achieved in a standard vibratory process. A centrifugal barrel machine is typically comprised of four barrels horizontally mounted between two main drive turrets, as is shown in Figure 2. As the two main turrets rotate, all barrels rotate in the opposite direction of the turret. The capacities of centrifugal barrel machines range from 0.5 cf to 12 cf of processing volume.

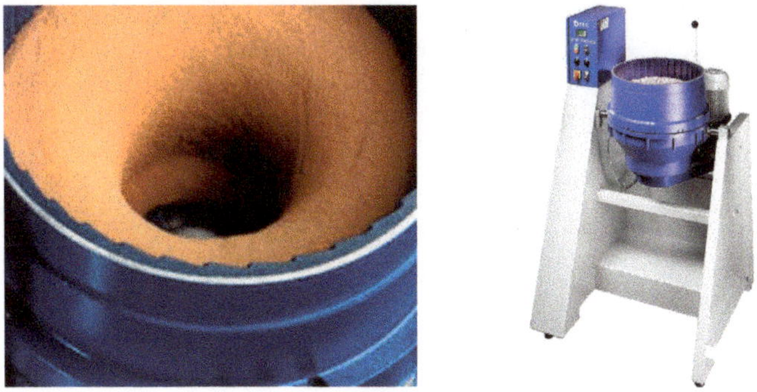

Figure 1. Centrifugal disc machine principle (LJMU).

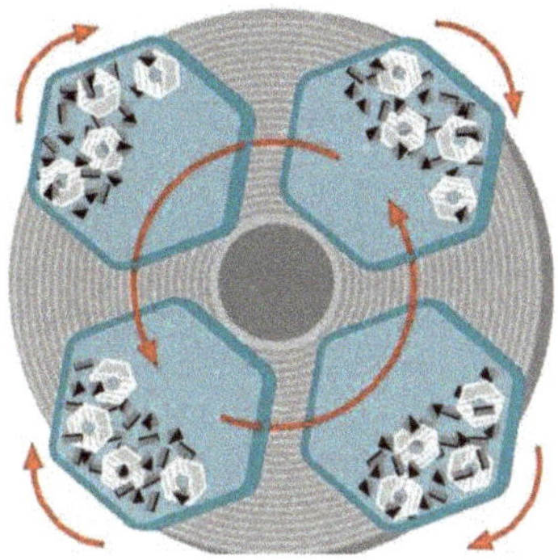

Figure 2. High-energy centrifugal barrel machine principle showing media and parts in each barrel.

This process requires manual loading and unloading as well as parts and media separation. Processes ranging from heavy grinding to fine polishing can be accomplished with centrifugal barrel finishing. Parts can be processed dry in a few select media types in a centrifugal barrel. Both wet or dry, a great deal of heat and pressure are generated within each barrel during this method. Heat must be considered from both a safety and finishing standpoint when centrifugal barrels are utilised.

2.1.3. Fixture Mass Finishing

The purpose of a fixturing method is to hold and give positional and/or rotational variation to the part and its interaction with the media. In the case of a fixed part, immersed in media and rotated at speed (e.g., drag finishing), the part edges and surfaces interact with loose media and a higher force is generated than that developed in conventional mass finishing processes, in which the part is placed randomly in the machine chamber and depends on the media motion to achieve the surface finishing result. Fixture arrangements also promote quicker cycle times. Furthermore, a benefit of the fixturing arrangement is that contact between parts is prevented. The best examples of fixture processes are drag finishing and stream finishing, which give high-quality results with short processing time and high reliability [11]. In this study, automation and optimisation of fixture finishing process were employed to help minimise human intervention.

Drag Finishing

Drag finishing machine technology is used for high-quality and sensitive workpieces that need to be ground or polished without contact between parts; it is available in dry and wet processing.

Workpieces are fixed in a special arrangement of workpiece holders. Holders are dragged in circular motion through the container that is filled with the loose abrasive or polishing media. The main spindle is also rotated at different speeds, either clockwise or anticlockwise. Figure 3 shows an image of the drag finishing machine principle.

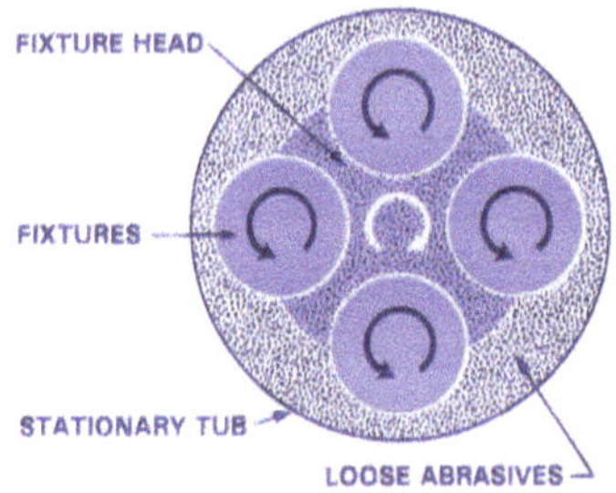

Figure 3. Drag finishing machine principle.

The mass finishing process is significantly dependent on pressure and speed. In the drag finishing process, the interaction between the high-speed workpiece and media will create high-pressure contacts, leading in a short time to the development of a high-quality surface finish and precise rounding of the part edges, which can otherwise only be obtained by the hand-polishing method. By controlling variables such as immersion depth and rotary speed, it is possible to reach rates of material removal up to 40 times higher than in conventional (vibratory, centrifuge) mass finishing and about 4–5 times faster than high-energy vibratory systems. In addition, the existence of a double drive system with two electric motors allows workpiece edges to be rounded uniformly. This also enhances the tool life. The applications for drag finishing range from aggressive deburring and edge breaking through to high-gloss polishing.

Stream Finishing

Stream finishing is a new method of mass finishing. It is also termed abrasion finishing, though it is not always used as an abrasion process but also for lapping and polishing, depending on the output target. It is able to finish workpieces of length up to 200 mm and weight of 20 kg, including the holder.

Figure 4 shows an image of the principle of a stream finishing machine. The workpiece holder is rotated either clockwise or anti clockwise to allow the workpiece itself to align to the direction of stream flow media. The media flow is generated through the bowl rotation at a specific speed. This configuration helps reduce the effect of transverse forces and lowers the risk of the tool breaking or becoming deformed [1]. Stream finishing is short in processing time, simple in operation, and has excellent reliability. However, despite the high processing force it can generate a fine surface roughness value (Ra) less than 0.03 μm.

The machine shown is able to process six workpieces in unison in a single holder fitted with a quick release mechanism to ensure that the workpiece is finished over the entire surface. Workpiece holders (examples: jaw chuck, drill chuck, collets and gripper) are available possessing different independent features that can rotate around their own axis at the same time. The processing time can be reduced significantly by adapting high-speed rotating workpiece holders (up to 2000 rpm).

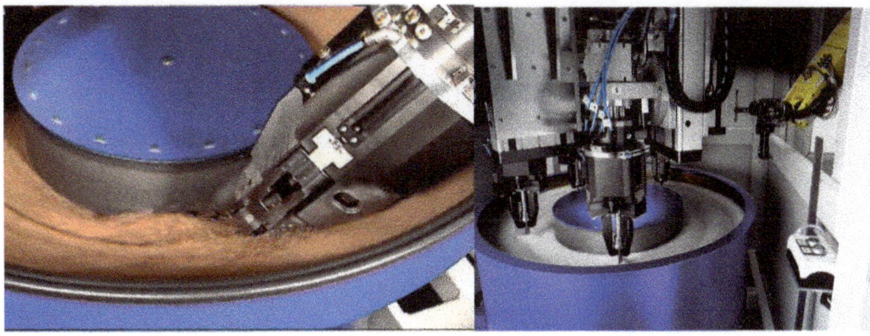

Figure 4. Stream finishing machine principle (LJMU).

2.2. Mass Finishing Media

There are five major materials of loose abrasive media (ceramic, plastic and synthetic, metal, precision, and micro-bite). The internal composition of conventional media is the source of the cutting action, weight, wear, and level of surface finish, all of which define the capabilities of the media.

2.2.1. Shape, Size and Weight

There are approximately 15 different major shapes of loose abrasive media led by triangular, conical, cylindrical, spherical and tetrahedral. Each abrasive media is used for a specific application depending on the workpiece configuration. Shape is an important factor in media selection. The major considerations are that the shape of the loose abrasive media chosen should permit access to all surfaces of the workpiece—meaning any ledges, holes, or other geometric features—for the deburring or finishing operation, and it must allow an easy separation of the media from the workpiece treated at the end of the cycle. In some cases, a mixture of media shapes may be most effective.

The size of loose abrasive media is important for several reasons. It helps to keep the workpieces separate from each other. The right choice can allow the media to reach all surfaces to finish or deburr.

Larger media have a faster cutting rate and produce smoother surfaces on workpieces. However, the use of small media in high-energy methods of mass finishing can improve the surface finish without increasing the cycle time. The weight of the loose abrasive media is another important factor in media selection.

According to [1], heavier media allow faster cutting. However, heavier media apply more pressure than lighter media, which can cause thin or ductile workpieces to bend or distort.

2.2.2. Media Capability

The capability is the ability of the media to deburr corners, edges and other surface requirements with uniformity (constant performance over time). The capability of a media is principally defined by: the machining parameters; the media wear rate; media composition (cutting action, weight, wear and surface finish); the ability to finish a range of workpieces (i.e., shape, size, material); and quality of the finishing. Hard media will work longer, but softer media will cut faster. In terms of media wear, most plastic, abrasive, or ceramic media get smaller over time, which can affect the finishing operation if they are lodged in holes and other recesses, which requires removing them manually.

The efficiency of material removal is strongly determined by the sharpness, hardness and density of the abrasive cutting edges. A media with an aggressive cutting surface will result in a poor surface finish and a media with a fine cutting efficiency will result in a good surface finish. This aspect is essential to understand the capability of the media assessed in this study.

The abrasive grain distribution on the conical shape media has been shown [1] to be largely independent to stock removed. A study directed by Moore [12] investigating the relationship between the volume of workpiece abrasive wear and the mean diameter of the abrasive particles shown that it is possible to find the most economical solution by determining the ratio of media to workload. The results of that study indicate that any change in the optimum ratio can produce a significant effect on the quality of finishing achieved [1]. It can also be noted that, for the range of materials used in the study, increasing the diameter of the abrasive grain increases the volume that is removed from the workpiece. Domblesk and Dennis [13,14] suggested that the particle shape is not the only variable that controls the behaviour of manufactured abrasives. The relationship between particle shape and packing behaviour needs to be understood before the former can be used to accurately predict the wear-rate.

2.3. *Additive Manufacturing (AM) Artefact Design for Finishing*

2.3.1. Design Constraints of Part to Be Finished

The geometry for the finishing artefact is influenced by three main factors:

- ➢ Design space: The artefact should be bound within a Ø40 mm × 80 mm design space. This was required so that the artefact could move freely within the high-energy machine, as shown in Figure 5.
- ➢ Selective Laser Melting (SLM) using Renishaw AM250. Design points to take into account:
    - Build direction
    - Minimum feature size (refer to design features)
    - Build supports
    - Size of part (SLM machine can build parts larger than this design space)
- ➢ Design features: The part had to include a variety of design features within a compact design space. Including: thru holes, free-form surfaces and sharp corners.

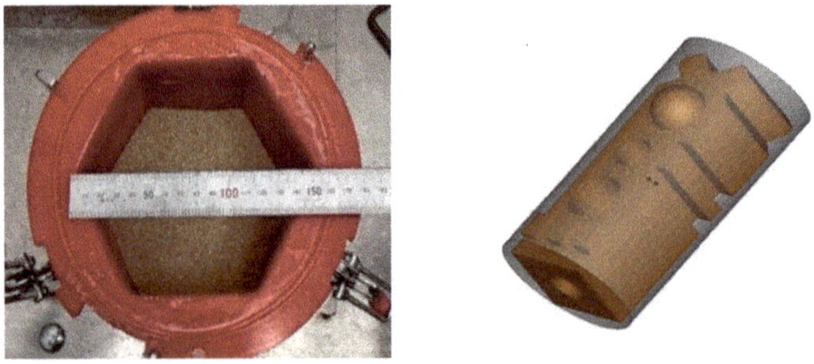

Figure 5. High-energy centrifuge and design space for part (grey shaded area).

2.3.2. Artefact Design Summary

- ➢ Part Size: 25 mm × 25 mm × 80 mm
- ➢ Design Features (Figure 6)
    - Free-form surface
    - 2 × Angled surfaces (45 degrees to horizontal)
    - Concave and convex features
    - 2 × cut out sections (5 mm × 5 mm × 25 mm)
    - 10 mm and 15 mm thru holes
    - 3 × blocks (1, 2 and 3 mm height)
    - Flat surface (25 mm × 65 mm)

➢ Fixturing point:

- 2 mm pilot hole for internal thread.

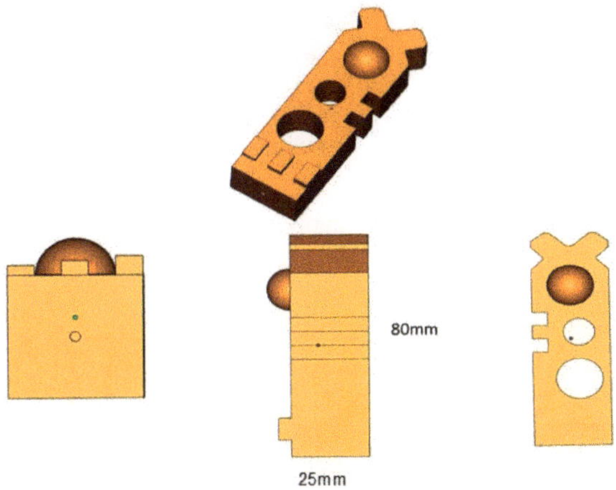

Figure 6. The resultant additive manufactured (AM) artefact design features.

2.4. AM Build Strategy

Build height: 32.5 mm
Build time: ~40 h
Supports: Minimal supports are required for this build
Machine: Renishaw AM250
Parts per build: Maximum of 10 parts per build

The base materials used in this study were Ti-6Al-4V. All parts were built using a Renishaw AM250 selective laser melting (SLM) machine. According to a complex part manufacturing setting, the vertical strategy is used to create a Ti thin section sample with layer thickness of 30 μm. The Ti-6Al-4V powder diameters used for experiments were located within the range of 10–20 μm and the composition conformed to the standard, Table 1. Argon gas is used during the process in order to protect the oxidation of surfaces. The gas is supplied through the powder projection nozzle of the additive manufacturing machine. The operating parameters used for the build samples are listed in Table 2. The average surface roughness of the base material is around 20 μm. The build configuration on the plate is shown in Figure 7.

Table 1. Chemical compositions for Ti-6Al-4V used in the study (wt %).

Al	V	O	N	C	H	Fe	Ti
5.5–6.75	3.5–4.5	0.2	0.05	0.08	0.015	0.3	Bal.

Table 2. Build parameters.

Build Parameter	Value
Power (W)	300
Beam size (μm)	85
Point distance (μm)	60
Effective speed (mm/s)	500
Hatch spacing (μm)	50

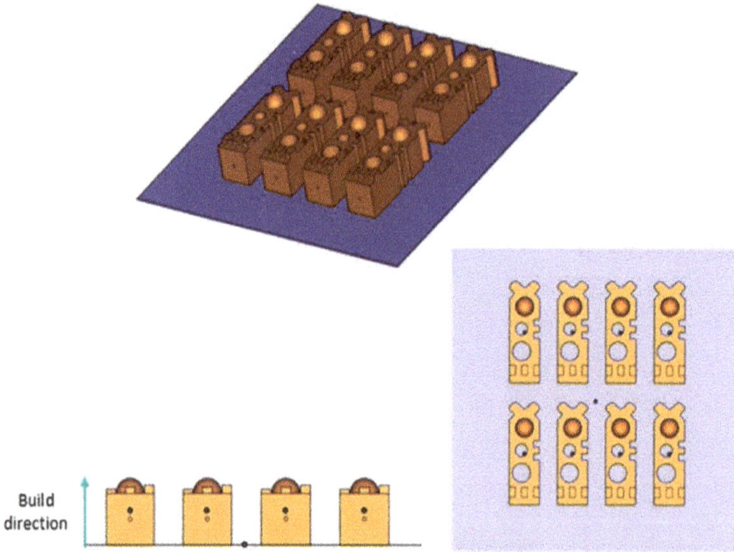

Figure 7. AM building strategy (showing eight parts).

2.5. Experimental Design and Setup

The experimental analyses were carried out on the comparison between different mass finishing processes of drag, stream, disc centrifuge and high-energy centrifuge. The AM metal parts in Stainless Steel 316 were used in this investigation with average surface roughness values, following build and sand blasting, of 1.7 μm (±0.15 μm). The liquid compound SC15 was used for the tests. The proposed techniques were compared according to the 2D surface finish parameters (Ra, Rz, and Rt) and areal surface finish parameters (Sa, Sz, and St) achievable at constant processing time intervals of 30, 60, 90 and 120 min. The experimental parameters are summarised in Table 3.

Three different media were used during the finishing process: plastic (OTEC DS, conical), ceramic (OTEC KM10 and KXMA24, tetrahedral) and Al Oxide (spherical). The experimental tests that involved drag finishing were performed by setting the head speed at 50 rpm, and varying the spindle speed over the values of 20, 40, and 80 rpm. In the case of stream finishing, the experimental tests were performed by setting the spindle speed at 30 rpm, and changing bowl speed over the values of 20, 40 and 60 rpm. The position of the workpiece in the bowl was selected carefully at the maximum stream flow region. In both cases full immersion of the workpieces were employed to ensure better performance results. The work holding design has been fabricated according to the workpiece geometry. The angled head was selected to be zero in all operation processes in order to ensure optimal contact between the media and workpiece.

The experimental tests relating to the centrifugal disc finishing, were performed by varying the rotating speed within the values of 150, 200 and 250 rpm, while the rotation speed kept at 200 rpm for the high-energy centrifuge.

The 2D surface roughness of the workpieces was inspected by tracing the surface using a Talysurf surface instrument at several positions along the surface before and after finishing processes (stress force = 0.75 mN; standard cut-off = 0.1 mm; transverse length = 0.8 mm; amplitude height 2.5 μm; stylus speed = 0.5 mm/s) Three surface roughness measurements were performed on the workpiece surface at each time interval; the average of the group results are presented in this study.

A three-dimensional optical profiler instrument was used to assess the morphology of the specimen before and after the machining process. The optical profile instrument (GFM) uses the phase measurement fringe projection technology method and is a good tool for characterising surface height variations and quantifying the amplitude of peaks and valleys (roughness profile) resulting from the machining process.

Table 3. Full range of machining parameters. HE: high-energy.

Drag Finishing	
Main head speed (rpm)	50
Spindle speed (rpm)	20, 40, 80
Processing time (min)	30, 60, 90, 120
Immersion depth (mm)	300/full immersion depth
Angled head (degree)	0
Media	Plastic KM10, Ceramic KXMA24
Compound	SC15
Dosing (%)	3.9
Stream Finishing	
Spindle speed (rpm)	30
Bowl speed (rpm)	20, 40, 60
Processing time (min)	30, 60, 90, 120
Immersion depth (mm)	300/full immersion depth
Angled head (degree)	0
Media	Plastic KM10, Ceramic KXMA24
Compound	SC15
Dosing (%)	3.9
Centrifugal Finishing	
Disc speed (rpm)	150, 200, 250
Processing time (min)	30, 60, 90, 120
Media	Ceramic KXMA24
Compound	SC15
Dosing (%)	1
HE Centrifugal	
Finishing	150, 200, 250
Disc speed (rpm)	30, 60, 90, 120
Processing time (min)	Ceramic
Media	SC15
Compound	Wet condition
Dosing (%)	1

3. Results

The results of the comparative evaluation between the four advanced finishing process techniques (drag, stream, disc, and high-energy (HE) centrifuge) are summarised in Figures 8–11, which provide the surface roughness values at different stages of the manufacturing process.

The results show that each of the finishing processes improved the starting workpiece morphology by progressively cutting the peaks and obtaining a smooth surface based on the setting of the operational parameters. The surface roughness profile (Figures 8 and 9) shows that the valley surfaces remain largely unaffected by the finishing process, this may be attributed to the media size, which cannot reach these regions except in stream finishing where fine spherical Al Oxide media was employed. It is clearly noticed, the relatively high spindle head speed of drag and stream, reduces the average surface roughness over the processing time regardless of the abrasive tool. However, the results demonstrated that the rate at which surface roughness decreases over the maximum cycle time is related to the AM part orientation. Table 4 shows the decreases in surface roughness using various mass finishing processes at maximum processing time of 120 min.

Table 4. Surface roughness values (average) of mass finishing processes at maximum processing time.

Finishing Process	Top Surface Ra (μm)	Bottom Surface Ra (μm)	Side Surface Ra (μm)	Angle Surface Ra (μm)
Drag	0.28	0.33	0.36	0.63
Stream	0.29	0.27	0.24	0.36
Disc	0.47	0.21	0.76	0.93
HE Centrifuge	0.48	0.39	0.42	0.75

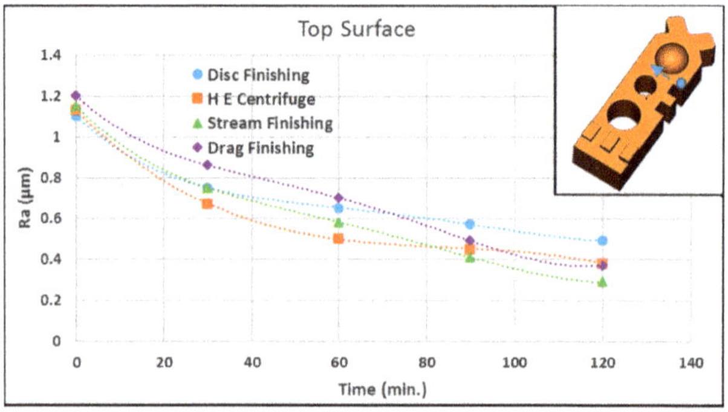

Figure 8. Surface roughness comparison on bottom surface (blue arrows show direction of measurement).

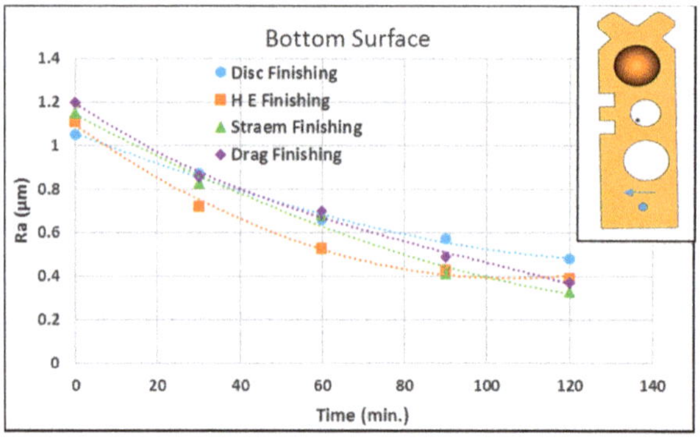

Figure 9. Surface roughness comparison on top surface (blue arrows show direction of measurement).

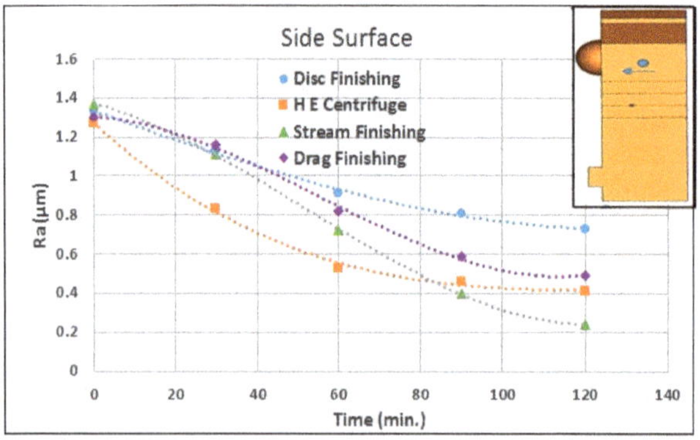

Figure 10. Surface roughness comparison on angle surface (blue arrows show direction of measurement).

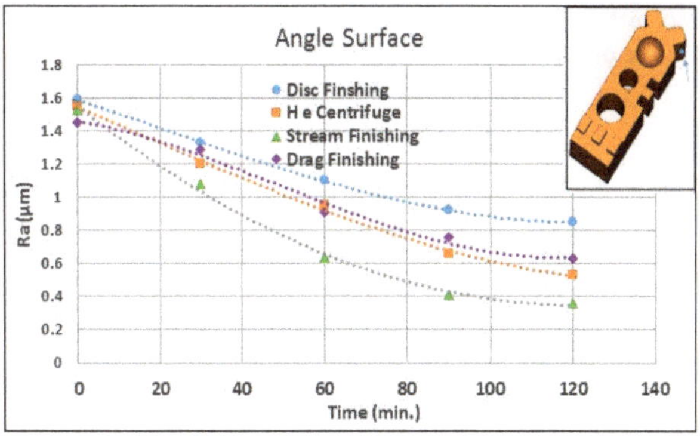

Figure 11. Surface roughness comparison on side surface (blue arrows show direction of measurement).

Mass finishing processes of the external surfaces, and easily accessible internal surfaces, to specification can be accomplished by traditional processes. However, most of these processes are inflexible and provide little control, nor can they access internal surfaces of complex shapes. For these type of challenges stream finishing can be a viable solution. Figure 12 shows the surface roughness (Ra) values of the internal hole surface as a function of processing time using the stream finishing process and Al Oxide media. It is clear that the surface roughness decreases to the value of 0.41 µm at the maximum processing time of 180 min.

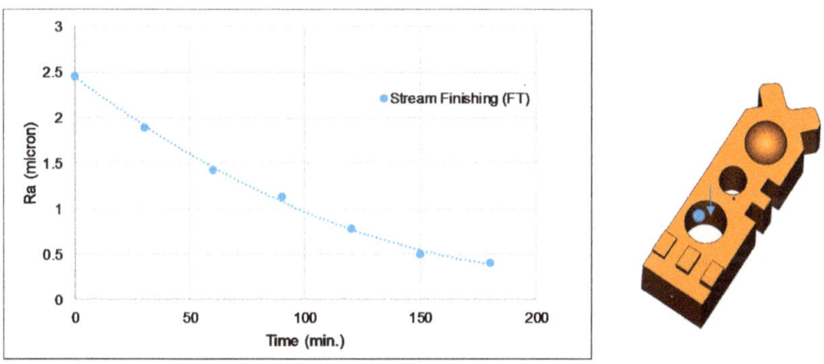

Figure 12. Stream finishing internal surface roughness values as a function of processing time (blue arrows show direction of measurement).

Figures 13 and 14 show the comparison of areal surface roughness (Sa, Sz, and St) of the side surface at maximum processing time of 120 min using different finishing technologies and media grade (ceramic media have been used for disc, HE centrifuge, and drag finishing processes and Al Oxide media for the stream finishing process). Figure 13 shows that the decreases in surface roughness Sa of the side surface using drag, stream, disc, and HE centrifuge at maximum processing time of 120 min were 3.2, 1.7, 2.3, and 1.3 µm, respectively. Similarly, in Figure 14 the HE centrifuge achieved the lowest Sz, and St values followed by stream, centrifuge and drag finishing processes.

Figure 15 shows an example of the 3D mapping assessment of AM part morphology using stream and HE centrifuge finishing processes. The 3D maps were obtained before and after 120 min of finishing. The results demonstrated that, after 180 min, the built layer peaks of the workpieces are very thin when processed with stream finishing and completely disappear with the HE centrifuge. This suggests that the HE centrifuge finishing process with ceramic media has the most aggressive uncontrolled effect on the AM part, which may cause over-finishing of the workpiece surface in the context of poor edge/radius retention, feature losses and breaching of tolerance criteria. This over-finishing may cause a failure of the workpiece due to the high surface erosion and may force a re-design.

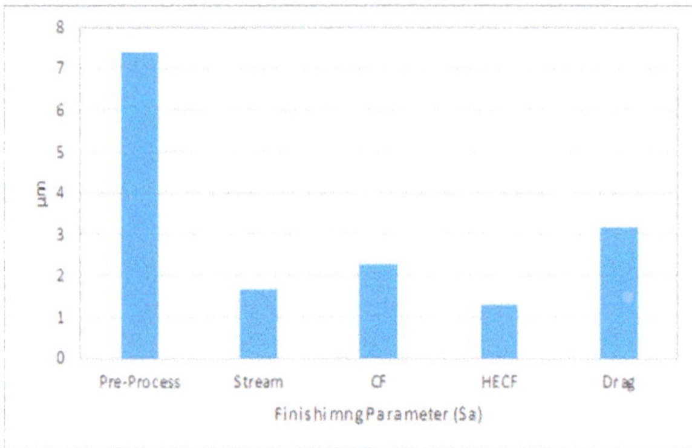

Figure 13. Sa comparison at maximum processing time of various mass finishing processes.

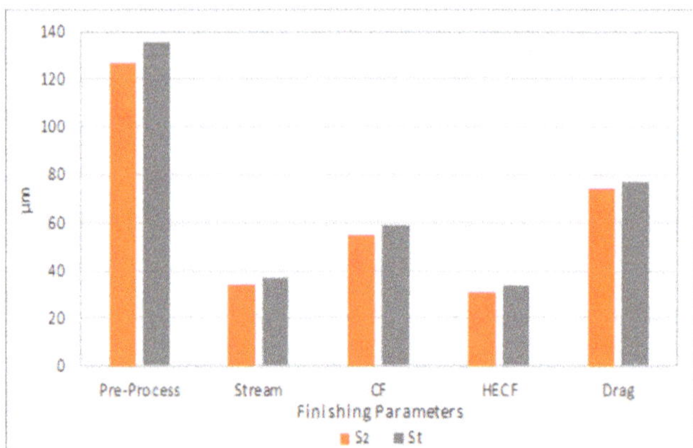

Figure 14. Sz and St comparison at maximum processing time of various mass finishing processes.

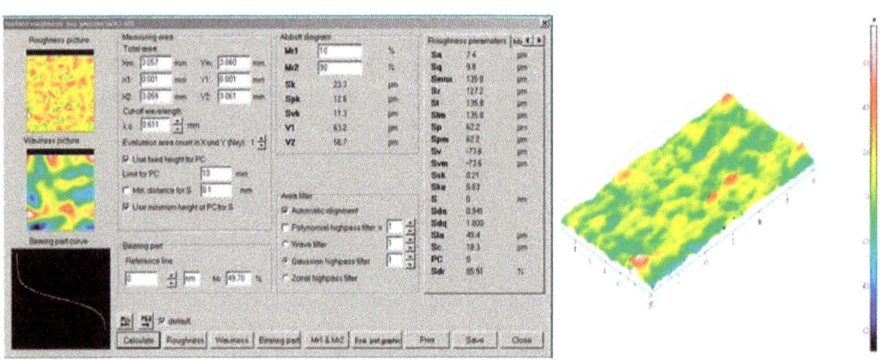

(a)

Figure 15. *Cont.*

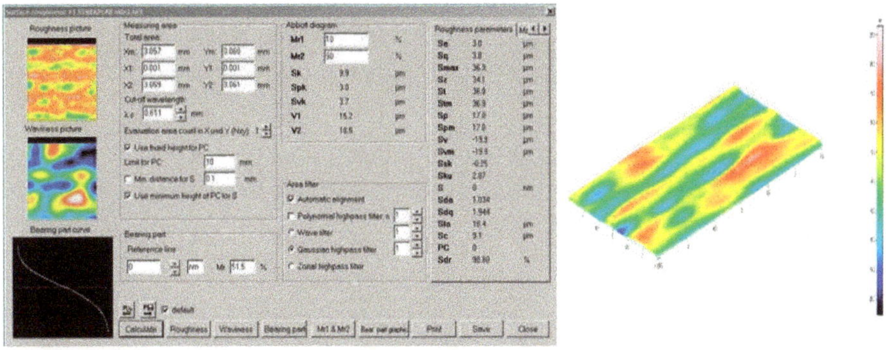

(**b**)

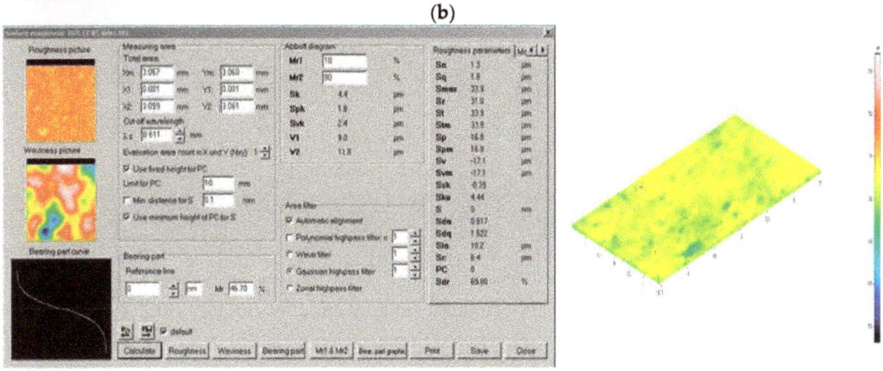

(**c**)

Figure 15. A 3D map assessment of AM part morphology before and after finishing processes. (**a**) Pre-processing 3D areal surface roughness parameters; (**b**) Stream finishing 3D areal surface roughness parameters; (**c**) HE centrifuge finishing 3D areal surface roughness parameters.

Despite the wide versatility of mass finishing processes, some potential process limitations should be noted. For the non-controllable finishing processes (disc and HE centrifuge), it can be difficult to selectively treat certain areas of the part to the exclusion of other areas, which might have critical dimensional tolerance requirements. Unless masked or fixtured, all exterior areas of the part will be affected by the process to a greater or lesser degree, with effects on corners and edges being more pronounced than those on flat areas, and with interior holes, channels, and recesses being relatively unaffected in the more common processes. However, care must be exercised in media size, shape selection, machining parameters and maintenance to prevent media lodging in holes and recesses, which might require labour-intensive manual removal and prohibit further abrasive/polishing action. Some parts have shapes, sizes, or weights that may prevent them from being finished in some traditional mass finishing processes because of the risk of impingement from part-on-part contact. However, drag and stream finishing processes can be a viable solution for such challenges.

4. Process Optimisation System of Drag Finishing

For this study, the core of the process optimisation system was based on readily available and well-understood proprietary statistical and mathematical software packages. The complete optimisation tool used ANOVA, the Taguchi experimental methodology and the response surface methodology (RSM) coupled with a MATLAB—Graphical User Interface (GUI). These tools were used to develop a surface roughness prediction model of advanced mass finishing technology under various machining parameters conditions. The model can then be used to estimate process parameters needed to achieve a desired roughness of a workpiece.

The usefulness of the system is not restricted to optimisation in the production context. It can also be used as design tool or as a production planning aid. As a design tool it can provide optimal parameter sets for materials new to the process or not in common use, and as a planning tool to plan production cycles to match related operations and to avoid misalignment of outputs with other processes, or to ensure maximum machine utilisation or indeed both. Further, it has potential to be of great benefit to machine tool salespersons and production engineers in the field.

4.1. The Taguchi Method

Taguchi is a statistical experimental tool designed to calculate a single or multiple groups of data (variable) to find the optimal and regularity in the operating environment (response). It is also used to identify the controllable and un-controllable factors in order to minimise the effect of noise factors. The orthogonal array (L8) matrix is designed to investigate the effect of each factor independently from others in order to reduce time and cost. The main objective is to identify the ranks of each factor that affect the mean response.

The statistical evaluation of performance required use of a signal-to-noise (S/N) ratio approach to measure deviation of the quality characteristic from the design value.

The S/N ratio can be used instead of the average value for the evaluation characteristic in the optimum variable analysis [15–17].

The S/N ratio approach can be divided into three quality tools: "the smaller the better", "the nominal the better", and "the higher the better" when characterizing the response quality for engineering processes. In this study, "the smaller the better" surface roughness was considered in investigating the influence of machining variables in the drag finishing process. The S/N ratio of "the smaller the better" quality characteristics can be defined by Equation (1):

$$\frac{\mathrm{S}}{\mathrm{N}} = -10\log\left[\sum_{i=1}^{n}\frac{Y^2}{n}\right], \tag{1}$$

where $(Y^2)/n$ is the mean of the square of measured data at i test and n is the number of repetitions.

Experimental Design and Setup

Experimental investigations were carried out using a drag OTEC machine. AM parts were used in the experiments with average surface roughness values of 1.7 μm (±0.15 μm). Liquid compound SC15 at a constant dosing rate of 3%. Ceramic and plastic media KM10 were employed in the tests. All experiments were performed at a constant processing time of 90 min.

The machining parameters were selected as follows: the experimental tests were performed with variable spindle speed, DSS within range ($10 \leq \mathrm{DSS} \leq 80$) rpm, the head speed, DHS various within range ($10 \leq \mathrm{DHS} \leq 60$).

Another important parameter that should be considered in the drag is the angle of the head holding the workpiece. Due to the shape of the AM part, the angled head was selected to be zero (vertical position) in all tests using a special holding arrangement to ensure continuous flow of media on the workpiece surface during operation.

Areal Sa surface roughness tests were conducted using a three-dimensional optical profiler instrument to investigate the morphology of the specimen before and after the machining process. Three surface roughness measurements were performed on each workpiece material, the average of the group results is presented in this study. However, this work can be extended to include the process optimisation system of Taguchi, ANOVA, and surface response methodologies to develop a mathematical model used to predict the surface roughness of the AM component, as shown in Figure 16. In general, the process optimisation system is based on 10 steps of planning, performing and evaluating results of a matrix of experiments to determine the performance levels of control parameters [18].

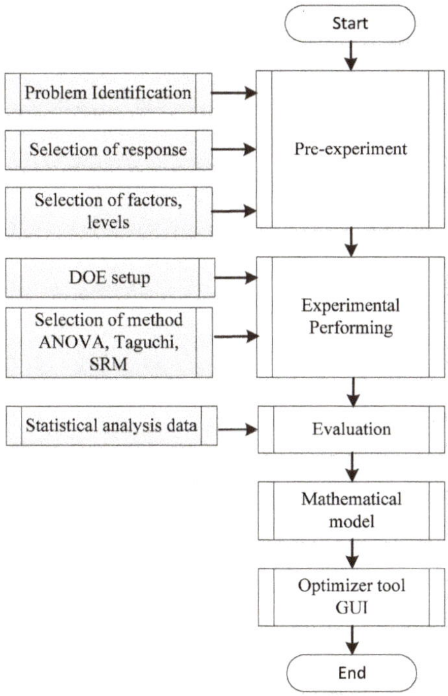

Figure 16. A schema of the process optimisation system.

4.2. Results

Analysis of the Signal-to-Noise (S/N) Ratio (Taguchi Method)

The S/N number represents the ratio between desirable and non-desirable values. The process parameter with the highest value has the most significant effect on the optimisation process [15]. In this study, the Taguchi optimisation process was selected with six degrees of freedom, three factors and two levels. The process parameters and their levels in the drag process are presented in Table 5.

Table 5. DOE layout and drag finishing results using Taguchi method.

Process Parameter	Level 1	Level 2
Spindle speed (DSS)	20	80
Head speed(DHS)	20	50
Media type	Ceramic	Plastic

The process variables and their levels were set over a wide range of parameters in order to cover a comprehensive performance capability of the finishing processes. A set of experiments were conducted using the dag finishing process and the L8 orthogonal array with eight experiments. Table 6 shows the DOE layout and results from the drag finishing processes.

Due to different units within the process parameters, each factor was coded with two levels. The influence of the interaction between variables is neglected in the Taguchi method. Regardless of the machining process characteristics, the greater the S/N value, the better the performance. Based on the S/N ratio analysis, the optimal drag finisher performance for best surface roughness was obtained at 80 rpm spindle speed (level 2), 50 rpm head speed (level 2), and Ceramic media (level 1).

Optimal finishing conditions for the drag process are shown in Table 7. Table 7 indicates that the strongest effect on the finishing process is given as rank 1 and is associated with the media, followed by rank 2 (the head speed), while rank 3 (the spindle speed) has the weakest effect on the process.

Table 6. Drag finishing process parameters and their levels.

No.	DSS	DHS	Media	Sa (μm)	S/N
1	1	1	1	2.0	−6.020
2	1	1	2	4.3	−12.669
3	1	2	1	1.9	−3.521
4	1	2	2	5.2	−10.370
5	2	1	1	2.5	−5.575
6	2	1	2	7.9	−14.320
7	2	2	1	1.5	−7.958
8	2	2	2	3.3	−17.952

Table 7. S/N ratios for surface roughness using drag finisher.

Level	DSS	DHS	Media
1	−9.646	−11.150	−5.769
2	−9.951	−8.447	13.828
Delta	0.305	2.704	8.509
Rank	3	2	1

Figure 17 shows the principal variations caused by changes to the S/N ratio. The strongest influence of the factor on the finishing process is measured by differences values. The factor that contributes to the higher difference in the mean S/N ratio over the selected design control factor, the more significant the impact. Optimal finishing conditions of design control factors of drag finishing process can then be determined from the S/N response graphs. As displayed, the media type shows the higher slope of the S/N ratio, which identifies the significant impact on the finishing process. The results show that the surface roughness decreases with the increasing drag head speed. The S/N ratios also identify that spindle speed had the least influence of the three factors.

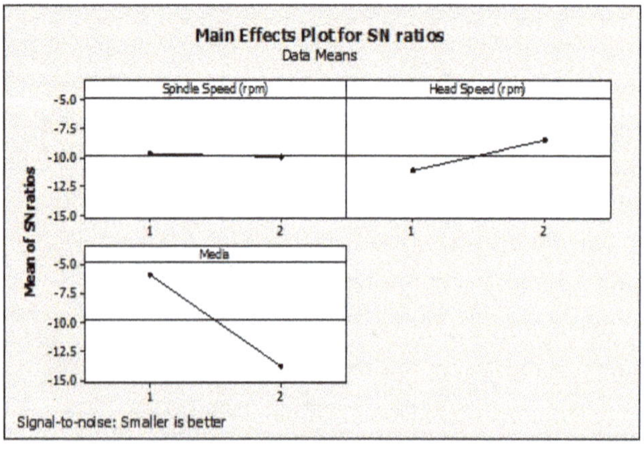

Figure 17. Effect of drag process parameters on surface roughness.

5. Conclusions

A comparison of results obtained using spindle finishing (drag and stream), centrifugal disc finishing and HE centrifuge processes using conventional abrasive granulate (Al Oxide and ceramic) media on the surface finish of an AM component has been undertaken. The results for the cases studied suggest that the stream finishing process is most efficient in reducing the surface roughness followed by HE centrifuge and drag finishing, then finally the centrifugal disc finishing process. However, it is to be noted that the media used in the drag machine was not the optimal type for this study.

The abrasion potential of the drag and stream finishing processes is high, especially when operated at the fastest spindle speed of 80 rpm and bowl speed 60 rpm, respectively. This may be attributed to the different hydrodynamic behaviour of the abrasive media in the processing medium. The impacts with abrasive media take place at high speed, which is generated from the difference between the workpiece rotating speed and the quasi-static abrasive media for the drag finishing process and from the difference between the workpiece rotation speed and abrasive media rotation flow for the stream finishing process. In the centrifugal finishing process, the workpiece is immersed into the abrasive media and travels with the flow.

Despite the high rotation speed of the disc in the centrifugal finishing process (250 rpm), the relative speeds between the abrasive media and the workpiece are expected to be lower because it only comes from the resultant difference in the system force acting on the abrasive medium. This speed is much higher in the HE centrifuge process, resulting in high-energy contact ,which alters the AM surface texture substantially, not only reducing surface peaks but generating a surface in which the positioning of the peaks has been altered appreciably.

However, it is the impact speed together with the mass of abrasive media that develop the kinetic energy that is released onto the workpiece surface. It follows that based on the abrasive media type, the abrasion potential of different technologies is determined by the impact speeds/force between the abrasive media and the workpiece. The energy of such interactions is converted to strain energy once impacted on the workpiece surface.

Furthermore, the strain energy release must be high enough in order to achieve a critical value equal to the resistance of crack growth, thereby leading to surface plastic deformations by micro-cutting. This higher energy should be accurately controlled to avoid problems such as the over-finishing of the workpiece surface. Over-finishing may cause a failure of the workpiece due to the massive surface erosion.

The reduced residual material after machining, low energy requirements, and quick cycle times make the fixed part processes of drag and stream industrially sustainable and promising in different manufacturing areas where high quality and performance are crucial requirements.

In this study, an experimental investigation has been undertaken with statistical analysis to determine the effect of varying DSS, DHS and media in the drag finishing process on the surface roughness values of AM components, using the Taguchi method. Taguchi analysis is responsible for investigating the rank and the level of factors that have a significant effect on the response. Taguchi statistical analysis of the drag finishing process suggested that the control factors that minimise surface roughness were ceramic media, DHS (level 2), 80 rpm and DSS (level 2), 50 rpm.

The study has identified the need to further characterise selected high-energy mass finishing processes for more realistic, and hence potentially more challenging, AM parts. The study has highlighted the potential for the future development of a two-stage finishing system.

Author Contributions: The research was conceived and completed jointly by Mikdam Jamal and Michael N. Morgan. The Manufacturing Technoogy Centre and partners, and Liverpool John Moores University technical staff assisted in the completion of experiments and surface measurements.

Conflicts of Interest: The authors declare no conflicts of interest.

References

1. Gillespie, L.K. *Mass Finishing Handbook*; Industrial Press: New York, NY, USA, 2006; ISBN 9780831132576.
2. Mediratta, R.; Ahluwalia, K.; Yeo, S.H. State-of-the-Art on Vibratory Finishing in the Aviation Industry: An Industrial and Academic Perspective. 2015. Available online: link.springer.com/article/10.1007/s00170-015-7942-0 (accessed on 1 July 2017).
3. Salvatorea, F.; Grangea, F.; Kaminskia, R.; Claudina, C.; Kermoucheb, G.; Recha, J.; Texierc, A. Experimental and Numerical Study of Media Action during Tribofinishing in the Case of SLM Titanium Parts. 2017. Available online: www.sciencedirect.com/science/article/pii/S2212827117304341 (accessed on 1 August 2017).
4. Walton, K.; Blunt, L.; Fleming, L. The Topographic Development and Areal Parametric Characterization of a Stratified Surface Polished by Mass Finishing. 2015. Available online: iopscience.iop.org/article/10.1088/2051-672X/3/3/035003 (accessed on 1 August 2017).
5. Zaki, N. Mass finishing considerations for optimum productivity. *Metal Finish.* **1992**, *90*, 50–54.
6. Brust, P.C. *Surface Improvement by Vibratory Cascade Finishing Process*; Transactions-North American Manufacturing Research Institution of SME: Lincoln, NE, USA, 1997; pp. 93–98.
7. Davidson, D.A. Developments in mass finishing technology. *Metal Finish.* **2003**, *101*, 49–56. [CrossRef]
8. Domblesky, J.; Evans, R.; Cariapa, V. Material removal model for vibratory finishing. *Int. J. Prod. Res.* **2004**, *42*, 1029–1041. [CrossRef]
9. Davidson, D.A. Surface finishing reaches new heights: Mass media finishing techniques can improve aircraft part performance and service life. *Metal Finish.* **2005**, *103*, 25–28. [CrossRef]
10. Cariapa, V.; Park, H.; Kim, J.; Cheng, C.; Evaristo, A. Development of a metal removal model using spherical ceramic media in a centrifugal disc mass finishing machine. *Int. J. Adv. Manuf. Technol.* **2008**, *39*, 92–106. [CrossRef]
11. Telljohann, D. Stream Finishing Process and Application. 2014. Available online: http://www.otec.de/ (accessed on 1 August 2017).
12. Moore, M.A.; Douthwaite, R.M. Plastic deformation below worn surfaces. *Metall. Trans. A* **1976**, *7*, 1833–1839. [CrossRef]
13. Domblesky, J.; Cariapa, V.; Evans, R. Investigation of vibratory bowl finishing. *Int. J. Prod. Res.* **2003**, *41*, 3943–3953. [CrossRef]
14. De Pellegrin, D.V.; Stachowiak, G.W. Assessing the role of particle shape and scale in abrasion using 'sharpness analysis': Part II. Technique evaluation. *Wear* **2002**, *253*, 1026–1034. [CrossRef]
15. Phadke, M.S.; Kackar, R.N.; Speeney, D.V.; Grieco, M.J. Off-line quality control in integrated circuit fabrication using experimental design. *Bell Syst. Tech. J.* **1983**, *62*, 1273–1309. [CrossRef]
16. Pignatiello, J.J., Jr. An overview of the strategy and tactics of Taguchi. *IIE Trans.* **1988**, *20*, 247–254. [CrossRef]
17. Pease, G. *Taguchi Methods: A Hands-On Approach to Quality Engineering*; Addison Wesley: New York, NY, USA, 1993.
18. Jamal, M.; Morgan, M.N.; Peavoy, D. A digital process optimisation, process design and process informatics system for high energy abrasive mass finishing. *Int. J. Adv. Manuf. Technol.* **2017**, *92*, 303–319. [CrossRef]

inventions

MDPI

Article

Direct Photonic Fusion of Vitrified Bonding Materials

Mark J. Jackson [1,2,*] and Martin J. Toward [1]

1 Department of Engineering, University of Liverpool, Liverpool L69 3BX, UK; mjtoward@liv.ac.uk
2 Bonded Abrasive Group, Cambridge, Massachusetts, MA 01239, USA
* Correspondence: bondedabrasive@gmail.com; Tel.: +1-785-342-4784

Received: 17 July 2017; Accepted: 16 August 2017; Published: 18 August 2017

Abstract: The purpose of this study is to show the effects of the direct fusion of raw materials used for vitrified grinding wheels by photonic interactions. The paper describes the construction of a sintering apparatus that employs a pulsed neodymium: yttrium aluminum garnet (Nd:YAG) laser to fuse a combination of raw materials such as ball clay, feldspar, and borax to form a partially-crystalline glass material. The experimental results show that lasers can replace traditional methods of glass frit formation by fusing raw materials used in the manufacture of glass bonds for vitrified grinding wheels. X-ray diffraction data shows that a glass with short range order has formed using the new method. The work described herein provides a new avenue for glass frit formation applied to grinding wheel manufacture.

Keywords: grinding wheels; materials; bonding; glass; glass–ceramic; laser; fusion

1. Introduction

When studying the structure of vitrified bonds (i.e., raw materials that vitrify and form a mechanical and chemical bond with abrasive grains), it is obvious that differences in bonding characteristics are common in grinding wheels made with the same batch formulation but made from a variety of commercially available raw materials. This is not surprising considering that clays and feldspar minerals vary in composition and have different concentrations of trace elements. Oxides present in clays and feldspars can affect crystallization behavior to a large degree [1]. The subsequent melting and devitrification behavior of these raw materials can greatly affect the products of fusion. One way to prepare grinding wheels is to use pre-fitted raw materials where oxide compositions can be tailored to suit a particular grinding operation [2]. The current method of pre-fritting is to melt clays and feldspars at high temperature in a specially prepared platinum crucible so that the molten mass can be quickly quenched from high temperature to prepare the melt for fritting [3–6]. This is done by pouring the molten glass into a suitable coolant to control the rate of glass formation prior to milling and crushing to form the powdered glass frit [7–10]. The main aim of this work is to increase the rate at which the bonding materials are fused to form a glass or glass–ceramic bond using a directed photon beam, which not only prepares the fused mass for fritting, but can be further developed to form a vitrified network in a grinding wheel without the need for fritting, thus reducing the time it takes to produce a commercially available grinding wheel.

2. Experimental Apparatus and Materials

A specially designed photonic sintering unit was constructed using a Rofin Sinar 90 W flash-lamp pumped, Q-switched yttrium aluminum garnet (Nd:YAG) laser (molecular vibrating) with a frequency range of 0–60 kHz (photon wavelength, $\lambda = 1.06$ μm (infra-red), photon frequency, $\nu = 2.8 \times 10^{14}$ Hz, quantum efficiency ~40%). The laser consisted of a standard cavity design with the active medium being Nd^{3+} ions in a YAG crystal rod mounted to one of the foci of an elliptical chamber made from gold-coated aluminum. At the opposing end was a krypton lamp whose emission spectrum suits the Nd^{3+} activity profile. Mode control was activated by a Q-switch that creates controlled photonic oscillations in such a way that a mechanical chopper, optoelectronic switch, or an acoustic-optic switch can be used to control the mode of the light beam [11]. For this particular laser, the acoustic-optic switch propagates sound waves through a crystal lattice generated by a piezo-electric transducer. A radio frequency oscillator generates sound waves in the 24–27 MHz range that produces waves with an alternating refractive index, causing it to act like an optical grating and stopping the action of the laser.

Approximately 60–100 W of power is required to deflect a 60 W beam, so the Q-switch has to be cooled during laser operation. The speed of the optical shutters allows energy to build up in the cavity when the laser action is halted, thus creating very high peak power to be generated. For a 20 W Nd-YAG laser, Q-switch action may produce 1 mJ/pulse (pulse widths ~6 ns), resulting in the generation of 100 kW peak power. The pulse energy and pulse widths used in the current work were in the range of 0.2–3 mJ and 80–250 ns, respectively. For pulse frequencies between 15 and 60 kHz, the output power ranged from 2–12 W (lamp currents ~8–20 A (Figure 1)).

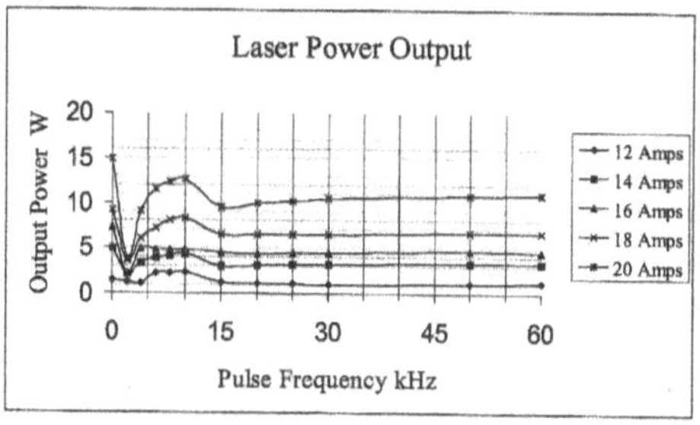

Figure 1. Laser power as a function of pulse frequency and lamp current for the Rofin Sinar 90 W flash-lamp, pumped, Q-switched yttrium aluminum garnet (Nd:YAG) laser.

Scan line overlap was ~50% of line width, and the peak power density was 1.4 GW/cm^2. Line scanning was achieved using a scanning head galvanometer containing two thermally regulated galvanometers giving a scanning speed range of 1–500 mm/s over an 80 mm × 80 mm scanning area. The focal length of the imaging lens was 112 mm, giving a minimum spot size of 50 μm. The experimental apparatus is shown in Figure 2a.

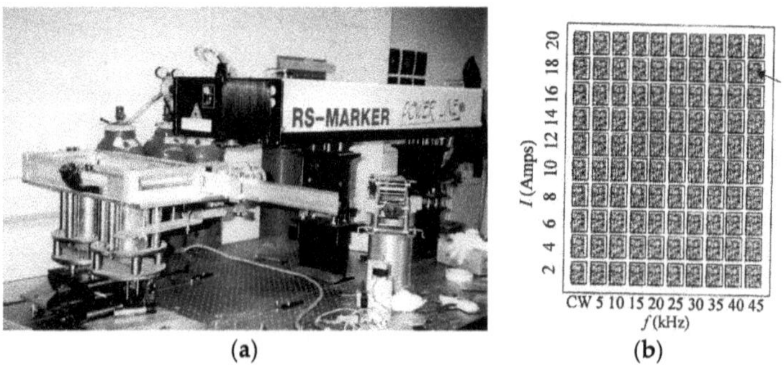

Figure 2. (**a**) Photonic sintering apparatus showing Q-switched Nd-YAG laser plus lens and scanning head galvanometer mounted directly above the build chamber; and (**b**) Schematic layout of laser–material interaction design-of-experiments showing magnitude of lamp current versus beam pulse frequency. Arrow indicates the first, or initial, location of interaction experiments.

The build chamber consists of a build cylinder and powder delivery cylinder, each of 100 mm diameter powered by two linear stepping motors with a minimum step size of 250 nm and maximum stepping rate of 380 steps/s. A thin layer of powder (10 mm thickness) was mounted on the build platform so that laser–material interactions could be characterized during the experimental procedure. The parametric variables investigated were frequency (0–60 Hz), scanning velocity (~50–500 mm/s), lamp current (8–20 A), focal length (104–112 mm), and spot size (100–300 μm). The system configuration enabled a matrix of 100 parametric values to be examined. The schematic layout of the laser–material interaction experiments is shown in Figure 2b.

The materials used in the experiments are those typically used in vitrified grinding wheels and include Devon ball clay (BC), Finnish flotation feldspar (FFF), borax frit (BX), and mixtures of all three raw materials. The preliminary results showed that no interactions occurred between the pulsed photonic beam and the ball clay up to 17.3 A, then vaporization occurred. Rapid heating of a powder results in a shock wave traveling towards the laser beam which vaporizes as it leaves the surface at velocities >10^5 m/s. However, on inspection of the ball clay, there was no evidence of powder blasting or cratering on the clay surface. This situation is tolerable because ball clay is used in grinding wheel manufacture to support abrasive particles together as a matrix while the fusible component of the bond melts and fills the interstices between clay particle and abrasive grain.

3. Experimental Results

The experimental work focused on the effects of changing process variables such as laser spot size, focal length, scanning velocity, lamp current, and optical frequency on the physical appearance and crystallographic structure of the fused bonding materials. The bonding materials processed were ball clay (BC), feldspar (FFF), borax frit (BX), and a mixture composed of 60 wt. % frit (BX), 20 wt. % ball clay (BC) and 20 wt. % feldspar (FFF).

Initially, X-ray diffraction of the raw materials was performed to characterize each bond ingredient using a Phillips 1710 X-ray generator with a 40 kV tube voltage and a 30 mA current. Monochromatic Cu kα radiation, λ = 1.54060 Å (wavelength intensity ratio ~0.5, divergence slit angle ~0.5°, receiving slit angle ~0.2°), was employed in the analysis. A scanning speed of 2° per minute for diffraction angles of 2θ between 15° and 60° was recorded by computer (start angle ~5° (2θ), end angle ~5° (2θ), step size ~0.04° (2θ)). The spectrum of each raw material was analyzed and compared with known spectra. Powder specimens were prepared by crushing in a mortar and pestle in preparation for quantitative X-ray diffraction. To eliminate the requirement of knowing mass absorption coefficients of ceramic samples for quantitative X-ray diffraction, Alexander and Klug [12] introduced the use of an internal standard for powder X-ray analysis. The powder bed subjected to "X-rays" should give the maximum diffracted intensity. The spectrum for ball clay (BC) is shown in Figure 3a. The spectrum shows kaolinite ($Al_2Si_2O_5(OH)_4$) at 2θ angles (plane) [*d*-value (α_1)] of 12.285° (001) [7.199 Å], 19.79° (020) [4.4826 Å], 24.815° (002) [3.5851 Å], 34.92° (-201) [2.5673 Å], 35.925° (200) [2.4978 Å], 37.65° (003) [2.3872 Å], 50.955° (004) [1.7907 Å], 59.855° (114) [1.544 Å], 63.895° (223) [1.4558 Å], 68.255° (062) [1.3730 Å], and 72.14° (204) [1.3083 Å]. The next major component of ball clay is quartz (SiO_2), and shows 2θ angles (plane) [*d*-value (α_1)] of 26.585° (101) [3.3503 Å], 40.25° (111) [2.2388 Å], 54.81° (202) [1.6736 Å], and 67.655° (212) [1.3837 Å]. Muscovite mica ($(K, Na)Al_2(SiAl)_4O_{10}$) also appears in the spectrum, but is of low measurable order.

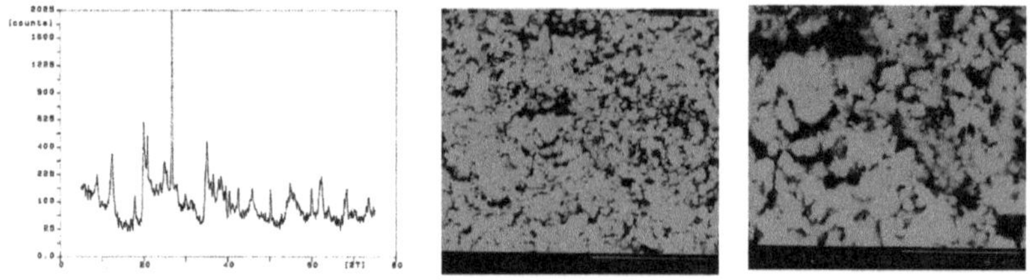

(a) (b) (c)

Figure 3. Ball clay (BC) material sample: (**a**) X-ray diffraction pattern for 2θ angles showing arbitrary counts (counts) versus 2θ angles (2T) in degrees; (**b**) Scanning electron micrograph at 500× magnification; and (**c**) Scanning electron micrograph at 1250× magnification.

The spectrum for Finnish Flotation Feldspar (FFF) is shown in Figure 4a. The major components detected include quartz (SiO_2) showing 2θ angles (plane) [*d*-value (α_1)] of 26.585° (101) [3.3503 Å], 40.25° (111) [2.2388 Å], 54.81° (202) [1.6736 Å] and 67.655° (212) [1.3837 Å] and albite ($NaAlSi_3O_8$) showing reflections at 020, 003, 112, 040 and 004 planes with corresponding *d*-values of 6.42, 4.24, 3.756, 3.21 and 3.18 Å, respectively.

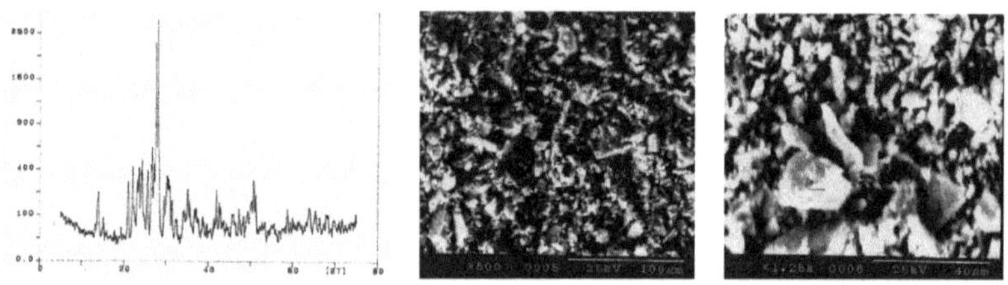

(a) (b) (c)

Figure 4. Finnish flotation feldspar (FFF) material sample: (**a**) X-ray diffraction pattern for 2θ angles showing arbitrary counts (counts) versus 2θ angles (2T) in degrees; (**b**) Scanning electron micrograph at 500× magnification; and (**c**) Scanning electron micrograph at 1250× magnification.

The spectrum for Borax frit (BX) is shown in Figure 5a. The major components include anhydrous borax ($Na_2B_4O_7$), which is approximately 31 wt. % Na_2O and 69 wt. % B_2O_3, and quartz (SiO_2) showing 2θ angles (plane) [*d*-value (α_1)] of 26.585° (101) [3.3503 Å], 40.25° (111) [2.2388 Å], 54.81° (202) [1.6736 Å], and 67.655° (212) [1.3837 Å].

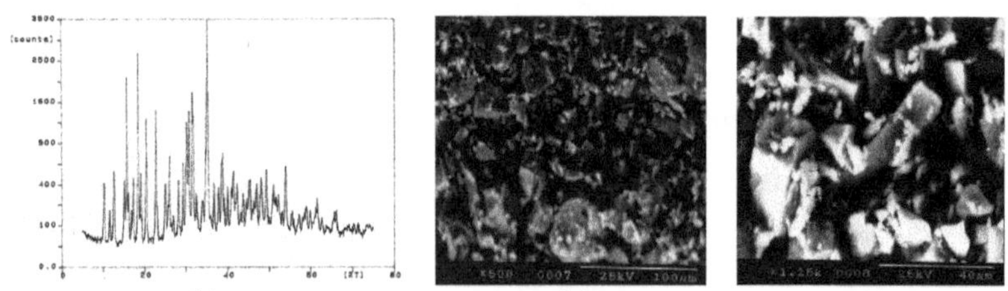

(a) (b) (c)

Figure 5. Borax frit (BX) material sample: (**a**) X-ray diffraction pattern for 2θ angles showing arbitrary counts (counts) versus 2θ angles (2T) in degrees; (**b**) Scanning electron micrograph at 500× magnification; and (**c**) Scanning electron micrograph at 1250× magnification.

The bonding materials were then handled as powders loaded manually in the build chamber to ensure direct photonic interactions between laser beam and powder bed as a single layer presented on a planar surface (Figure 6). The raw materials were exposed to the photonic beam, which projected itself as a spot of various sizes (100, 200, and 300 μm) scanned across the powder bed at various speeds—typically 50, 250, and 500 mm/s.

Figure 6. The build chamber showing a thin layer of powder presented as a single layer on a planar surface prior to direct photonic interaction.

The following observations were made for each material after a thin layer was exposed to the laser beam at various spot sizes and laser scan velocities:

(i) Ball clay (BC)—for a 100 μm spot size and 50 mm/s scan velocity, there was no interaction between laser and ball clay up to 9.8 A lamp current, beyond 9.8 A and at a frequency of 26 kHz, burning of the ball clay occurred. When the scan velocity was increased to 250 mm/s, vaporization began at 18.6 A lamp current between 0 and 60 kHz. Beyond 500 mm/s scan velocity, no interaction was observed between the photonic beam and the ball clay. For 200 μm spot size, slight burning of the ball clay powder occurred at 16.9 A lamp current between 0 and 60 kHz frequency; at larger spot size (~300 μm), no interactions occurred at higher scan velocities, although burning occurred at 50 mm/s scan speed at 16.9 A lamp current between the 0 and 60 kHz frequency range. The topography of the BC powder bed is shown in Figure 3b,c at 500× and 1250× magnifications, respectively.

(ii) Finnish floatation feldspar (FFF)—at 100 μm spot size and 50 mm/s scan velocity, discoloration of the powder occurred up to 10.6 A lamp current between 0 and 60 kHz frequency, after which vaporization occurred. Above 33 kHz frequency, the powder appeared to crystallize. This observation occurred at higher scan velocity until a 20 A lamp current produced complete crystallinity. With large spot sizes, crystalline phases were created at lamp currents >18.6 A. For a 300 μm spot size, FFF crystallinity occurred above 17.8 A lamp current between 0–60 kHz frequency range. For higher scan velocities (>250 mm/s), FFF became white in color. The topography of the FFF powder bed is shown in Figure 4b,c at 500× and 1250× magnifications, respectively.

(iii) Borax frit (BX)—no interactions were observed up to 11.9 A lamp current through the frequency range of 0–60 kHz. Beyond 11.9 A lamp current, slight fusion of the frit occurred at low scan velocities. The topography of the BX powder bed is shown in Figure 5b,c at 500× and 1250× magnifications, respectively.

(iv) Sintered bond mix—at 100 μm spot size and 50 mm/s scan velocity, no interactions were observed up to 9.8 A lamp current in the 0–60 kHz frequency range. Between 9.8 and 11.9 A lamp current, vaporization occurred with slight crystallinity observed in the bond mix. At lamp currents beyond 13.8 A, burning of the mix occurred. At higher scan velocities, partial fusion or no interactions occurred. Similar observations were made with large spot sizes, the optimum conditions being lamp currents between 9.8 and 11.9 A in the 0–60 kHz frequency range to produce vaporization of the mix accompanied by slight crystallinity. The topography of the bond mix powder bed is shown in Figure 7b,c at 500× and 1250× magnifications, respectively. The expected equilibrium phases from the directly fused mixture (20 wt. % BC, 20 wt. % FFF, and 60 wt. % BX) are quartz (unreacted and partially dissolved), mullite, cristobalite, and glass. However, from the samples tested, the compounds quartz, mullite, and glass were successfully detected, indicating non-equilibrium formation conditions

(Figure 7a). A calibration curve was constructed using a suitable internal standard (CaF_2), a diluent, and a synthetic form of the phase(s) to be measured. Synthetic mullite with purity >99.8% and powdered quartz with purity >99.84% SiO_2 were used throughout the experiments. The method used for quantitative analysis of ceramic powders was developed by Khandelwal and Cook [13] and refined by Monshi and Messer [14].

The internal standard provides an intense (111) reflection (d = 1.354 Å) lying between the (100) reflection for quartz (d = 4.257 Å) and the (200) reflection for mullite (d = 3.773 Å). Using copper kα radiation (λ = 1.5405 Å), the corresponding values of diffraction angle 2θ are: (100) quartz = 20.82°; (111) calcium fluoride = 28.3°; and (200) mullite = 32.26°. The calibration curve generated by varying proportions of calcium fluoride, synthetic quartz, and mullite is shown in Reference [8]. Mass fractions of the crystalline phases in the mixture can be interpreted from the calibration lines by measuring the intensity ratio of the phase(s) to the internal standard. The diffraction peaks of interest for quantitative analysis lie between 15° and 40° of the diffraction angle 2θ [8]. The reflections are the (111) planes of calcium fluoride, (200) planes of mullite, and the (100) planes of quartz. In order to calculate the mass fractions of quartz and mullite in the mixture, the height of the chosen diffraction peak and its width at half-height were measured from the diffraction spectrum. The product of these two measures were then compared with that of the internal standard, and the resultant intensity ratio was used to find the exact mass fraction of the phase(s) measured in the glass that was subjected to X-ray diffraction [8].

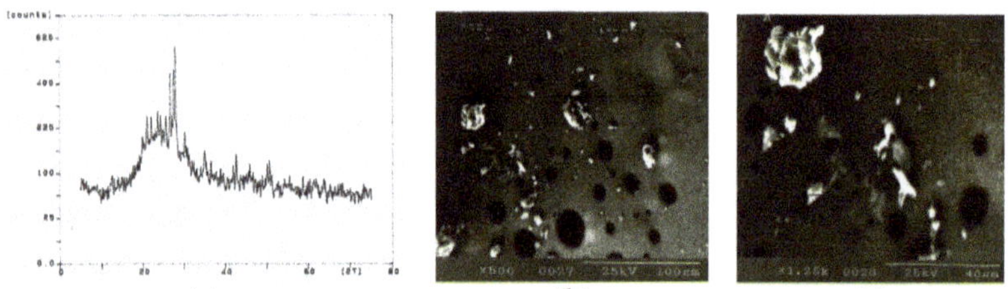

(a) (b) (c)

Figure 7. Sintered bond mix (20 wt. % BC, 20 wt. % FFF, and 60 wt. % BX) material sample: (**a**) X-ray diffraction pattern for 2θ angles showing arbitrary counts (counts) versus 2θ angles (2T) in degrees; (**b**) Scanning electron micrograph at 500× magnification; and (**c**) scanning electron image at 1250× magnification.

The X-ray spectrum for bond mixture is shown in Figure 7a. The major components detected include quartz (SiO_2) showing 2θ angles (plane) [d-value (α_1)] of 26.585° (101) [3.3503 Å], 40.25° (111) [2.2388 Å], 54.81° (202) [1.6736 Å], and 67.655° (212) [1.3837 Å], mullite ($Al_6Si_2O_{13}$) showing reflections at (110), (200), (001), (111), and (201) planes with corresponding d-values of 5.39, 3.774, 2.886, 2.542, and 2.292 Å, respectively, and a non-crystalline glass phase. The images shown in Figure 7b,c show that the surrounding bond mix has melted, while large clay relicts have remained intact. The bubbles shown are decomposed gases held in the clays that form and escape from their matrix during melting. These gases are typically CO_2 and SO_2, and are seen in molten glass frits during standard grinding wheel manufacture. The release of gases contained in clays and various pore forming agents occurs at high temperatures (>650 °C). The images shown in Figure 7 and the X-ray spectra of the bond mix subjected to laser beam interactions are not dissimilar to that of a glass frit made using the traditional way of melting where bonding materials are placed in a platinum crucible, heated to the melting temperature, then poured in a cooling furnace for the preparation of fritting prior to use in vitrified grinding wheels.

4. Discussion

The fusing of the raw materials was apparent at low scanning speeds and spot sizes. The powder formed a brittle structure that was full of porosity. The depth of fusion was ~0.5 mm for the bond mix. However, ball clay (BC) scanned separately burned at low scan speeds and spot sizes. At higher levels, no interaction between the photonic beam and the clay occurred. This is advantageous for the preparation of frits for grinding wheels because the clay particles remain intact as the lower melting point material flows around it. It is noted that the melting point of ball clay (BC) is ~1400 °C. Feldspar (FFF) particles behaved a little differently. Slight melting occurred at the edges of the feldspar particles at low scan speeds and spot sizes (FFF melting point is ~1200 °C). Although some melting is encouraged for feldspar, complete melting is not required in the bond mix because a large size grinding wheel may collapse under its own weight when fired due to the melting of too many bond components at the same temperature. The most surprising of all the raw materials was the behavior of borax frit (BX) during photonic interactions. The melting point of borax frit is ~742 °C, so one should have anticipated melting during the photonic interactions. However, direct interaction with BX frit tends to indicate that the photon beam is reflected by the thin powder layer of the BX frit. When mixed with the BC and FFF, BX frit appears to melt and form a highly non-crystalline structure that is shown in the X-ray spectrum shown in Figure 7a. This tends to indicate that the heat absorbed by BC and FFF by exposure to the photonic beam may be responsible for heating and causing BX to melt when presented as a mixture to the laser beam. This presents a very interesting situation in terms of providing raw materials for vitrified grinding wheels, because materials that reflect optical energy may be blended with materials that absorb energy in order to provide homogeneous melting of the bond mixture that bonds to abrasive grains. This implies that abrasive grains may not require heating during grinding wheel manufacture that uses lasers, thus preserving the initial strength and mechanical characteristics of abrasive grains that are normally impaired due to long periods of traditional thermal treatment in kilns.

Future research may therefore lead to the development of grinding wheels whose abrasive grains maintain their original properties due to the absence of thermal degradation caused by the use of focused photonic energy to selectively heat parts of the grinding wheel during manufacture. The avoidance of using clays altogether may be possible for small grinding wheels, because clays may be replaced by oxides whose composition may be tailored to suit the particular grinding application. However, the fusibility of oxides must be examined to ensure that photonic interactions can truly achieve the production of glass frits without the use of clays. It is expected that experiments described in this paper will lead to the development of new inventions associated with the production of glass bonding systems and new innovations in the way that direct photonic fusion can be used in the manufacture of vitrified grinding wheels. New inventions will be focused on producing vitrified grinding wheels that vitrify in a shorter amount of time using much less energy than current methods. The current work has shown that not only is direct photonic fusion of vitrified bond materials possible, but when applied to grinding wheel manufacture, the process may preserve the fracture characteristics of virgin abrasive grains without the deleterious effects of thermal degradation caused by traditional methods of grinding wheel manufacture.

Acknowledgments: The authors thank Paul Dando of Saint-Gobain Abrasives for supplying the raw materials and John Curran for training Martin Toward to use the laser sintering apparatus and its construction. The authors also thank Professor Bernard Hon for allowing the authors to conduct the work at the Product Innovation Development Center, University of Liverpool. The authors thank the Engineering and Physical Science Research Council for funding this work under grant number GR/N06731/01.

Author Contributions: Mark J. Jackson and Martin J. Toward conceived and designed the experiments; Martin J. Toward performed the experiments; Mark J. Jackson and Martin J. Toward analyzed the data; and Mark J. Jackson wrote the paper.

Conflicts of Interest: The authors declare no conflict of interest.

References

1. Jackson, M.J.; Hitchiner, M.P. Vitrified Bonding Systems and Heat Treatment. In *High Performance Grinding Advanced Cutting Tool*; Springer Nature: New York, NY, USA, 2013; pp. 45–94.
2. Graf, W. Overview of Abrasives. In *Handbook of Creep-Feed and Surface Grinding*; Winterthur Schlieftechnik AG: Winterthur, Switzerland, 2010; p. 21.
3. Jackson, M.J.; Mills, B. Interfacial bonding between corundum and glass. *J. Mater. Sci. Lett.* **2000**, *19*, 915–917. [CrossRef]
4. Jackson, M.J. Studies on refractory bonding systems used in vitrified silicon carbide grinding wheels. *Proc. Inst. Mech. Eng.* **2000**, *214*, 211–221. [CrossRef]
5. Jackson, M.J.; Mills, B. Vitrification heat treatment and the dissolution of quartz in grinding wheel bonding systems. *J. Inst. Mater. Miner. Min.* **2001**, *100*, 1–8. [CrossRef]
6. Jackson, M.J.; Wakefield, R.T.; Jones, S.A.; Mills, B.; Rowe, W.B. Materials selection applied to vitrified corundum grinding wheels. *J. Inst. Mater. Miner. Min.* **2001**, *100*, 229–236.
7. Jackson, M.J.; Mills, B. Microscale wear of vitrified abrasive materials. *J. Mater. Sci.* **2004**, *39*, 2131–2143. [CrossRef]
8. Jackson, M.J. Tribological design of grinding wheels using X-ray diffraction techniques. *Proc. Inst. Mech. Eng.* **2006**, *220*, 1–17. [CrossRef]
9. Jackson, M.J. Sintering and vitrification heat treatment of cBN grinding wheels. *J. Mater. Proc. Technol.* **2007**, *191*, 232–234. [CrossRef]
10. Jackson, M.J. An historical review of wear mechanisms and the structure of vitrified grinding wheels. *Int. J. Nanomanuf.* **2009**, *3*, 368–397. [CrossRef]
11. Steen, W.M. Chapter 1—Background and General Applications. In *Laser Materials Processing*, 2nd ed.; Springer: Berlin, Germany, 1998; pp. 11–57.
12. Alexander, L.; Klug, H.P. Basic aspects of X-ray absorption. *Anal. Chem.* **1948**, *20*, 886–889. [CrossRef]
13. Khandelwal, S.K.; Cook, R.L. Effect of alumina additions on crystalline constituents and fired properties of electrical porcelain. *Am. Ceram. Soc. Bull.* **1970**, *49*, 522–526.
14. Monshi, A.; Messer, P.F. Ratio of slopes method for quantitative X-ray diffraction analysis. *J. Mater. Sci.* **1991**, *26*, 3623–3627. [CrossRef]

inventions

MDPI

Article

Grinding Fluid Jet Characteristics and Their Effect on a Gear Profile Grinding Process

Philip Geilert [1,*], Carsten Heinzel [1,2] and André Wagner [1]

[1] Stiftung Institut für Werkstofftechnik (IWT), Badgasteiner Strasse 3, D-28359 Bremen, Germany; heinzel@iwt.uni-bremen.de (C.H.); wagner@iwt-bremen.de (A.W.)

[2] MAPEX Center for Materials and Processes, University of Bremen, Bibliothekstrasse 1, D-28359 Bremen, Germany

* Correspondence: geilert@iwt-bremen.de; Tel.: +49-421-218-51184

Received: 26 September 2017; Accepted: 20 October 2017; Published: 25 October 2017

Abstract: Profile gear grinding is characterized by a high level of achievable process performance and workpiece quality. However, the wide contact length between the workpiece and the grinding wheel is disadvantageous for the fluid supply to the contact zone and leads to the risk of locally burning the workpiece surface. For the reduction of both the thermal load and the risk of thermo-mechanical damage, the usage of a grinding fluid needs to be investigated and optimized. For this purpose, different kinds of grinding fluid nozzles were tested, which provide different grinding fluid jet characteristics. Through a specific design of the nozzles, it is possible to control the fluid flow inside the nozzle. It was found that this internal fluid flow directly influences the breakup of the coolant fluid jet. There are three groups of jet breakup ("droplet", "wave & droplet", and "atomization"). The first experimental results show that the influence of the jet breakup on the process performance is significant. The "wave & droplet" jet breakup can achieve a high process performance, in contrast to the "atomization" jet breakup. It can therefore be assumed that the wetting of the grinding wheel by the grinding fluid jet is significantly influenced by the jet breakup.

Keywords: profile gear grinding; grinding fluid; grinding fluid nozzle; jet breakup

1. Introduction and the State of the Art

In grinding processes in general, a high risk of thermo-mechanical damage exists due to the kinematics of the abrasive grain in the contact zone as well as the large contact area between the grinding wheel and the workpiece. This can lead to residual tensile stresses and changes in hardness, which can have a negative effect on the lifetime of the component [1,2]. Especially in profile gear grinding processes, the large contact area between grinding wheel and workpiece makes it difficult to supply the grinding fluid to the contact zone. To avoid thermo-mechanical damage, in most cases, the supply of a grinding fluid is indispensable and typical in industrial applications. Due to a lack of experience in appropriately supplying grinding fluid in practice, high grinding fluid flow rates are normally chosen to achieve high process reliability.

Experience has shown that the productivity of the process, workpiece quality, and tool wear can be influenced not only by the chemical composition of the grinding fluid, but also by the grinding fluid supply. In addition, large grinding fluid reservoirs are needed when high grinding fluid flow rates are used, because of the necessary cooling and settling phases of the grinding fluid. Furthermore, high grinding fluid flow rates and grinding fluid pressures result in a high level of aerosol pollution, necessitating an exhaust system to ensure the health of the machine operator [3]. Therefore, the aim should be to reduce the grinding fluid volume without affecting the workpiece quality and the material removal rate. Thus, a high performance grinding fluid supply becomes particularly important.

Consequently, a grinding fluid supply should be developed on the basis of "as much as necessary—as little as possible" [4–7].

A nozzle contour that is adapted to the profile gear grinding wheel contour can considerably increase the efficiency of the grinding fluid supply. In addition, the jet velocity and the coherence of the jet are important in grinding fluid nozzle design. To break through the air belt surrounding the grinding wheel, an adapted jet velocity is necessary. When the circumferential speed of the grinding wheel and the velocity of the grinding fluid jet are nearly identical, the wetting of the grinding wheel increases. In contrast to widening jets, a coherent jet carries less air into the grinding contact zone, and a larger amount of grinding fluid flows through the grinding contact zone [8]. As a result of these adjustments, the grinding fluid flow rate can be significantly decreased without any negative effect on the workpiece surface.

In addition, the nozzle also influences the jet characteristics and the jet breakup in particular. While studies have been conducted in the research fields of "atomization technology" and "combustion engines", focusing on the nozzles and their jet breakup behavior, no research work is known that deals with grinding fluid nozzles and their influence on both jet breakup and the grinding processes.

2. Research Approach and Objective

The aim of the presented work is to study the influence of the jet breakup on the grinding process. Therefore, the relationship between the nozzle designs and the jet breakup characteristics is analysed. Dimensionless numbers like the Ohnesorge number and the Reynolds number are used to classify the nozzles with regard to their jet characteristics. In a second step, it is attempted to establish a relationship between these theoretical investigations and the jet breakup characteristics, as well as the influence on the grinding process. Therefore, a gear profile grinding process is used and the nozzles are compared with regard to their influence on a thermo-mechanical damage of the workpiece. This should provide a comprehensive understanding of the relationship between the grinding fluid nozzle design, the jet breakup, and the grinding process.

The gained scientific knowledge should allow grinding fluid nozzles to be designed for an optimal wetting of the contact zone in profile grinding. As a result, the performance and energy efficiency of gear grinding processes can be expected to increase.

3. Materials and Methods

In the following, the theoretical backgrounds for designing grinding fluid nozzles, as well as the different nozzles used for the experimental investigations are described. This is followed by a description of the methods used to characterize a fluid jet. Furthermore, the machine and workpiece used for the grinding experiments are specified, and the micromagnetic test for thermo-mechanical damage is also introduced.

3.1. Nozzle Designs

A jet, which can normally be described by a cylindrical volume, decays because of the instability that is driven by capillary forces. This is known, amongst other things, as the Plateau-Rayleigh-Instability. This effect explains how a jet decays into a particle chain as it progresses and breaks apart until it takes the form with the smallest surface energy, the droplet [9].

In this context, three dimensionless numbers are important to characterize a fluid jet. In particular, the jet breakup can be characterized in advance in the Ohnesorge vs. Reynolds number diagram. The jet breakup can be divided into three fields ("droplet" (a), "wave & droplet" (b), and "atomization" (c)) (Figure 1) [9,10]. Due to the high jet velocity, the fields (b)] and (c) are of importance for grinding fluid nozzles.

The Reynolds number (Re) is a dimensionless number that is used to predict the transition from laminar to turbulent flow.

$$Re = \frac{\rho \cdot v \cdot D_H}{\eta}$$
$$\rho : \text{oil density } [\text{g/cm}^3] \; v : \text{mean velocity} [\text{m/s}]$$
$$D_H : \text{hydraulic diameter} [\text{mm}] \; \eta : \text{oil dynamilc viscosity } [\text{Pa} \cdot \text{s}] \quad (1)$$

The Ohnesorge number (Oh) relates the viscous forces to the inertial and surface tension forces:

$$Oh = \frac{\eta}{\sqrt{D_H \cdot \rho \cdot \sigma}} = \frac{\sqrt{We}}{Re}$$
$$\sigma : \text{oil surface tension } [\text{kg/s}^2] \quad (2)$$

Thereby, it provides information about the characterization of the fluid atomization. The lower the Ohnesorge number, the weaker the friction losses due to viscous forces. This means that most of the inserted energy converts into surface tension energy and a droplet can be formed. The higher the Ohnesorge number, the more dominant is the internal viscous dissipation and the droplet decays into smaller particles [9,10].

The Ohnesorge number can be expressed as the square root of the Weber number divided by the Reynolds number (Equation (2). The Weber number is described in terms of the inertia force, F_ρ, divided by the surface tension, F_σ:

$$We = \frac{F_\rho}{F_\sigma} = \frac{\rho \cdot v^2 \cdot D_H}{\sigma} \quad (3)$$

By means of the Weber number, it can be evaluated how far the real shape of a droplet differs from a spherical shape. Thereby, it is a measure for the deformation of a droplet, which increases with an increasing Weber number. Therefore, the droplet will breakup into more and more tiny droplets with an increasing Weber number [11].

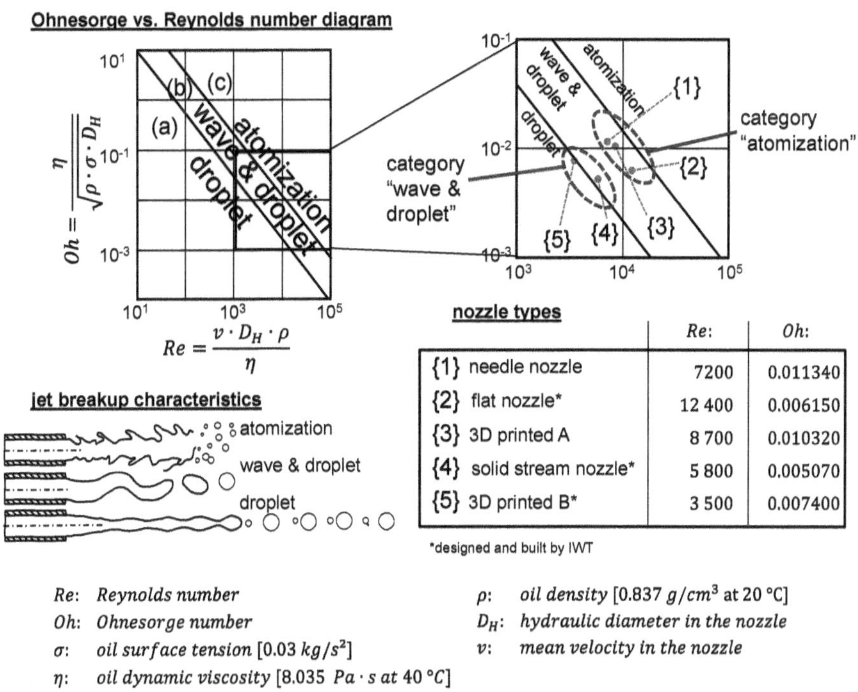

nozzle types	Re:	Oh:
{1} needle nozzle	7200	0.011340
{2} flat nozzle*	12 400	0.006150
{3} 3D printed A	8 700	0.010320
{4} solid stream nozzle*	5 800	0.005070
{5} 3D printed B*	3 500	0.007400

Figure 1. Ohnesorge vs. Reynolds number diagram and jet breakup characteristics (after [10]).

Based on this, grinding fluid nozzles {1–5} were chosen, which have an identical outlet cross-section and therefore provide the same average jet velocities (v_{jet} = 35 m/s) at the outlet for the same flow rates (Q_f = 100 l/min) (Figure 2). For comparison, the reference nozzle {r} was also investigated. This nozzle represents the current state of technology in the industrial environment. This nozzle differs from the nozzles {1–5} in the grinding fluid flowrate (Q_f = 330 l/min) and in the jet velocity (v_{jet} = 12 m/s). Due to the low jet velocity, the jet does not break up before it reaches the grinding contact zone.

To determine the positions for the nozzles {1–5} in the Ohnesorge vs. Reynolds number diagram and the characteristics of the jet breakup, the dimensionless numbers are calculated. The oil used has a surface tension of $\sigma = 0.03$ kg/s^2, a dynamic viscosity of $\eta = 8.035$ Pa·s, and a density of $\rho = 0.837$ g/cm^3. Using the mean velocity and the hydraulic diameter in the nozzle, which are specific to each nozzle, the Ohnesorge and the Reynolds numbers are calculated. The results are shown in Figure 1.

For these settings, the nozzles can be classified into three categories. The needle nozzle {1}, the flat nozzle {2}, and the three-dimensional (3D) printed nozzle A {3}, which are close to the area "atomization" in the diagram (category "atomization"). The solid stream nozzle {4} and the 3D printed nozzle B {5} are in the area "wave & droplet", close to the area "droplet" (category "wave & droplet"). The reference nozzle {r} generates a coherent jet with no breakup for the considered conditions (category "coherent jet") at a rather small jet velocity far below cutting speed v_c.

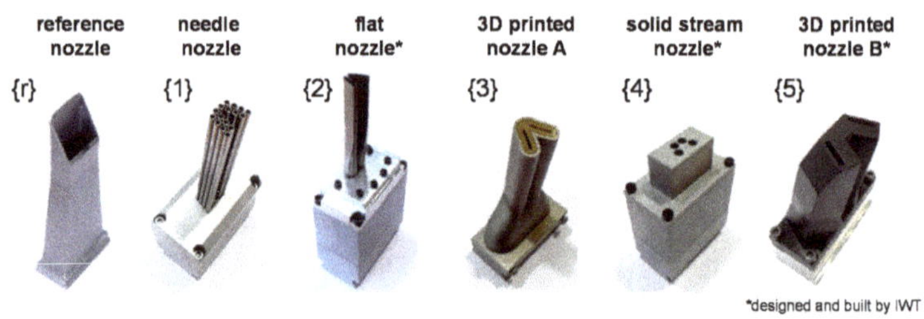

Figure 2. Different concepts of fluid supply nozzles.

3.2. *Characterization of Fluid Dynamic Aspects of Grinding Fluid Jets*

The grinding fluid jet breakup was characterized with the help of high speed photography. A camera was used to take high resolution pictures (20.2 mega pixel) with an extremely short exposure time. The pictures were illuminated from behind the grinding fluid jet, whereby a flash illuminated a translucent glass plate with an exposure time of 1/60,000 s. The even illumination of the background is transmitted through the grinding fluid jet. The pictures with the transmitted light through the grinding fluid jet allow knowledge to be gained regarding the jet characteristics, namely the widening of the jet, the distribution of the droplets, and the jet breakup (Figure 3).

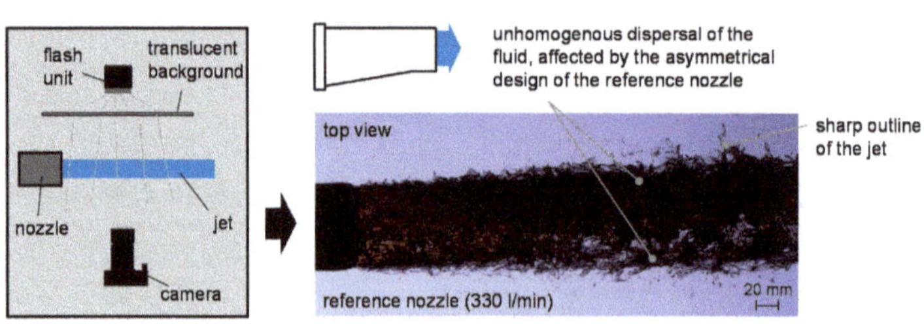

Figure 3. Measurement setup for high-speed photography.

The impact pressure of the grinding fluid jet was measured using a high resolution piezoelectric sensor (measuring tip diameter: 1.0 mm). The distribution of pressure was determined by moving the sensor through the grinding fluid jet (Figure 4, bottom right). A flow-optimized cover (Figure 4, top right) redirected the jet behind the measurement area, so that no deflected fluid could influence the measurement. These measurements allow for a better understanding of the distribution and the size of the droplets inside the jet. Thus it is possible to review the category of jet breakup for the nozzles.

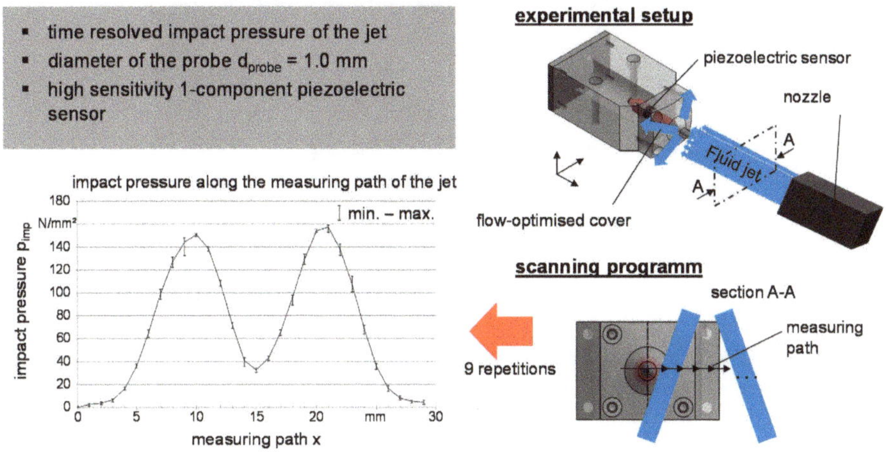

Figure 4. Experimental setup, scanning program and repeatability for impact pressure measurement.

The sensor scanned the grinding fluid jet ten times on one line and measured the force at several measuring points. The error bars indicate the min–max values (Figure 4, bottom left) and show reproducible results. In addition to the distribution of pressure, the time sequences were analyzed and it was possible to obtain knowledge about the regularity and the height of the impact pressure of the droplet chain.

The energy efficiency of the nozzles is evaluated with the static pressure within the pipe directly in front of the nozzle. The pressure is measured with a calibrated digital pressure transmitter. The used silicon sensor has a resolution of 0.1 bar.

3.3. Grinding Process

In the profile grinding experiments, pre-machined case-hardened gears were machined on a gear grinding machine (KAPP KX 500 FLEX), which enabled the machining of gears with a diameter of up to 500 mm as well as a module from m = 0.5 to 10 mm with conventional and superabrasive tools (Figure 5).

The workpieces are helical gears. These gears are characterized by 47 teeth, a normal module of $m_n = 4.5$ mm, a bevel angle of $\beta = -16.55°$, a pressure angle of $\alpha = 24°$, and a width of b = 65 mm. The material is AISI 5120 and the gears are case-hardened and blast-cleaned. The gears have a hardness of 718 HV and a case hardness depth Chd of 1.13 mm.

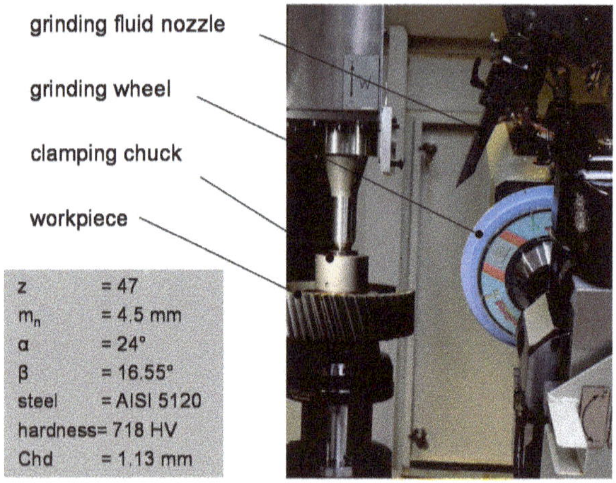

Figure 5. Experimental setup for grinding tests.

For the micromagnetic test to characterize the thermo-mechanical impact on the ground tooth flanks, the Stresstech measuring device Rollscan R 300 and the Stresstech NC unit GearScan 500 were used (Figure 6). These instruments enable a fully automatic testing of the gear. Hereby, a sensor moves by means of NC-controlled axes on previously programmed tracks along the tooth flank. Barkhausen noise was analyzed at the center of the right tooth profile. This automatic measurement allows a high reproducibility in comparison to manual tests.

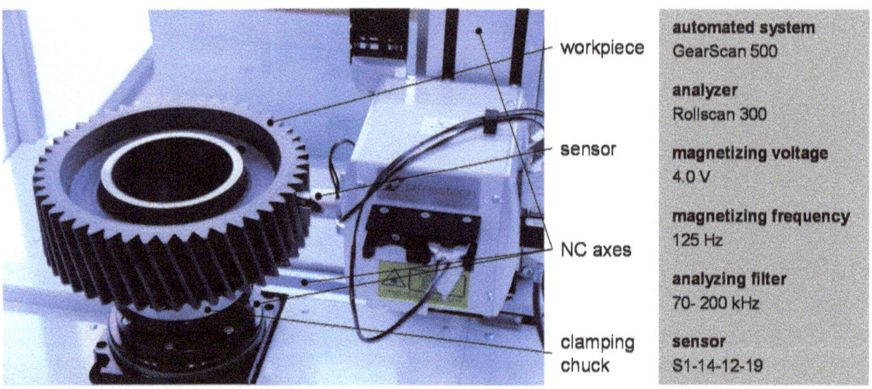

Figure 6. GearScan 500 in use and measurement parameters.

4. Results and Discussion

As explained below, the fluid dynamic investigations provide a characterization of fluid jets and their breakup. Furthermore, the grinding experiments show the impact of the jet characteristics on the grinding process.

4.1. Characterization of Fluid Jets and Their Breakup

Through the use of the above-described high-speed photography, the fluid jets of the different nozzles were characterized. A standardized measurement of the length of the coherent part of the jet

$l_{coherent}$ before atomization makes it possible to compare the fluid jets. The results for these lengths, which are the average lengths of nine measurements of each nozzle, are shown in Figure 7.

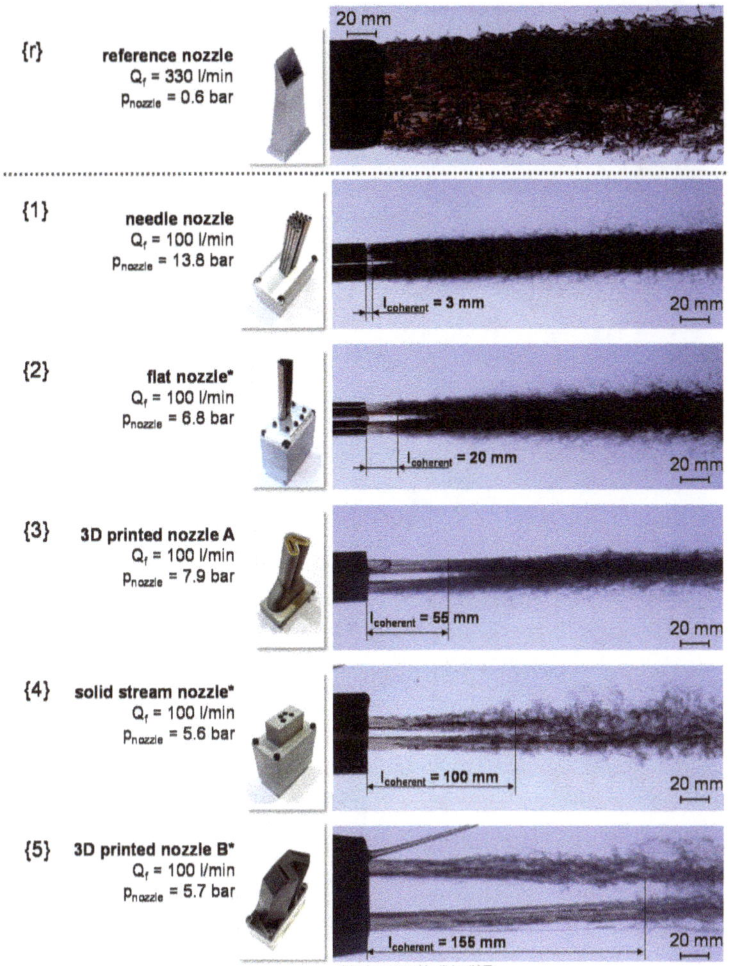

Figure 7. Characterization of fluid supply nozzles—coherent jet lengths.

The jets of the needle nozzle {1}, the flat nozzle {2}, and the 3D printed nozzle A {3} decay within a short length. The three values for the coherent length are similar to each other, as well as the values in the Ohnesorge vs. Reynolds number diagram are closely matched. The dark areas in the photos show many tiny droplets, which adsorb the light of the illuminated background. This observation points out that the three nozzles are comparable with regard to their position in the Ohnesorge vs. Reynolds number diagram and generate an "atomization" jet breakup. This might be due to the high turbulence directly at the nozzle outlets, which is caused by the high mean velocity inside the nozzles. The solid stream nozzle {4} and the 3D printed nozzle B {5} have inner designs with optimized flow characteristics, which significantly reduce the turbulence at the nozzle outlet. This can explain the results for the longer coherent lengths of these two nozzles. For the evaluation of the coherent lengths, only the respectively lower outlet was analysed, as defects in the 3D-printing of the upper outlet were determined for the 3D printed nozzle B {5} (these defects result in a short coherent length for the upper outlet). The results again reflect the relationship between the positions in the Re-Oh-diagram, which are comparable for both nozzles {4} and {5}, and the coherent jet lengths for this category.

Apart the jet characteristics, the nozzles influence the measured pressure in the pipe directly in front of the nozzle for a constant flow rate (Figure 7). For the needle nozzle {1} a pressure of p_{nozzle} = 13.8 bar is measured, whereby for 3D printed nozzle B {5} a pressure of p_{nozzle} = 5.7 bar is measured. The nozzle {2}–{4} are between these values. These pressure values correlate in inverse manner with the coherent lengths. These different power drops in the nozzles generate a corresponding turbulence flow, which correlates with the jet break up.

It can be summarized that the five nozzles {1–5} can be classified into the two categories "atomization" and "wave & droplet", and the results for the coherent jet lengths matches with this observation. For the category "wave & droplet" greater values for the coherent length can be reached than for the category "atomization".

Besides the optical jet characteristics, the fluid jets of the different nozzles differ significantly from each other with regard to the temporal course of the impact pressure. The impact pressure was measured for a nozzle of each category. The results of the measurement for the reference nozzle {r} (category "coherent jet"), as well as for the needle nozzle {1} (category "atomization") and the solid stream nozzle {4} (category "wave & droplet") are shown in Figure 8.

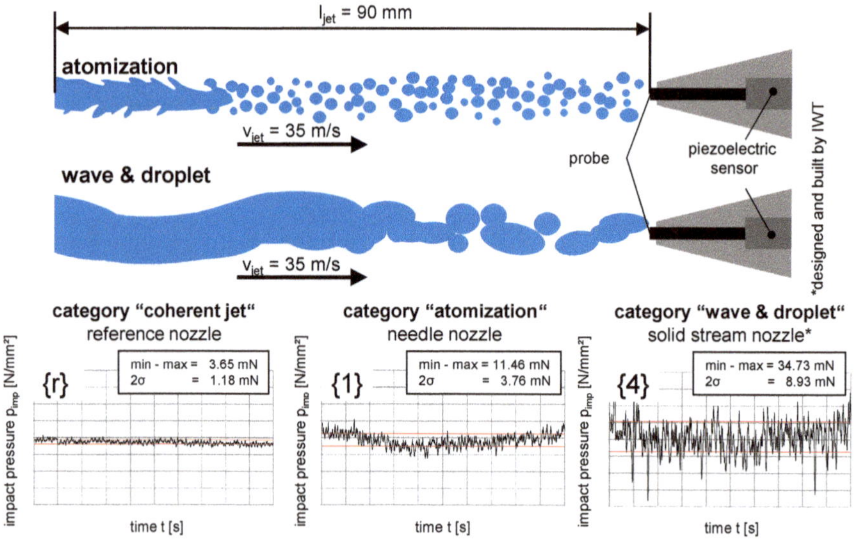

Figure 8. Characterization of the fluctuations in impact pressure for different jet breakups.

The reference nozzle {r} shows the lowest fluctuations in impact pressure. This is due to the low jet velocity and the coherent jet. Despite the equal and constant fluid flow rate and the jet velocity of the solid stream nozzle {4}, as well as the needle nozzle {1}, the solid stream nozzle {4} shows the most significant fluctuations in impact pressure. This can be explained by the distribution of the droplets, which is also shown in high-speed photos. The solid stream nozzle {4} generates a comparably long coherent grinding fluid jet that decays in the transitional area into relatively large droplets, which generates a greater impact pressure fluctuations than small ones. The fluid jet of the needle nozzle {1} shows an "atomization" breakup of the jet close to the nozzle outlet. The resulting small droplets are distributed evenly and therefore generate only minimal fluctuations, as well as a relatively low impact pressure.

The results show that grinding fluid nozzles not only have an impact on the widening of the jet as a function of jet velocity and flow rate, but also affect the jet breakup, the distribution of the droplets, and therefore the fluctuations in impact pressure.

4.2. Grinding Technology

Different grinding fluid nozzles were applied that show different jet characteristics ("coherent jet", "wave & droplet" and "atomization"). In order to compare these types of nozzles, a grinding process was developed that was tested three times with the reference nozzle {r} to determine the reproducibility. For a practical process and a fast reaching of the process limit, the gears were machined without pre-grinding in eight strokes and with a depth of cut a_e of 50 μm in each case. The dressing conditions were precisely adjusted to the requirements, so that an increased thermo-mechanical load of the surface layer could be achieved after grinding only a few gear gaps. The results for this reference process are shown in Figure 9.

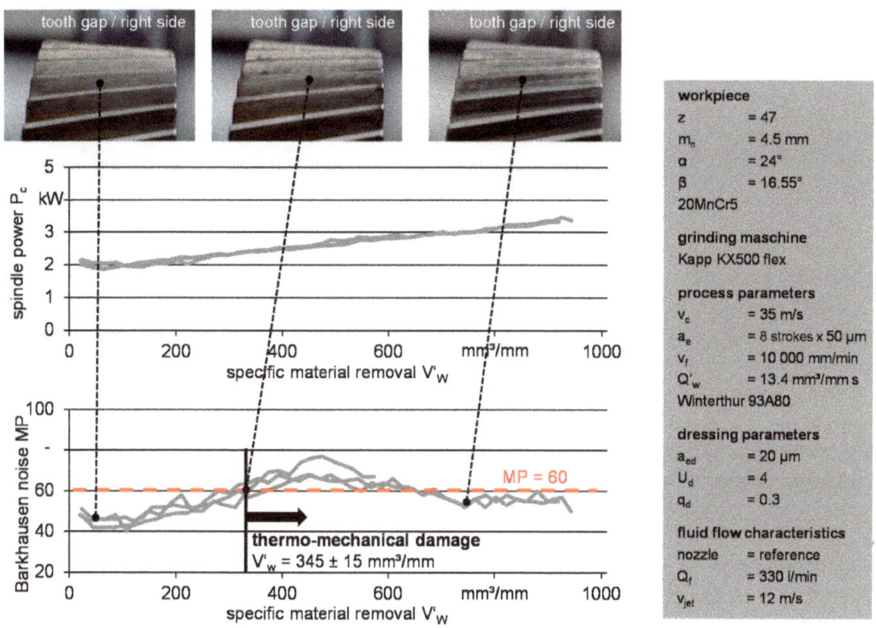

Figure 9. Barkhausen noise and spindle power for the reference gear grinding process.

The courses of the spindle power in all three of the repetitions are similar to each other, and the courses of the Barkhausen noise analysis are also closely matched. Hence, the repetition trials verify the reproducibility of the process. In order to compare the grinding fluid supply conditions for the six nozzles, the limit for the thermo-mechanical damage were compared using Barkhausen noise. Therefore, the limit for the thermo-mechanical damage for the reference process was determined with nital etching. When the nital etching indicates grinding burn, the Barkhausen noise value MP exceeds 60. For the reference process, a specific removed material volume V'_W of 345 ± 15 mm^3/mm was removed before this limit had been reached.

Based on the above-described grinding process, as well as the method to detect thermo-mechanical damage, all discussed nozzles were analyzed. The values for the specific removed material volume V'_W, before thermo-mechanical damage occurs, for each nozzle are shown in Figure 10.

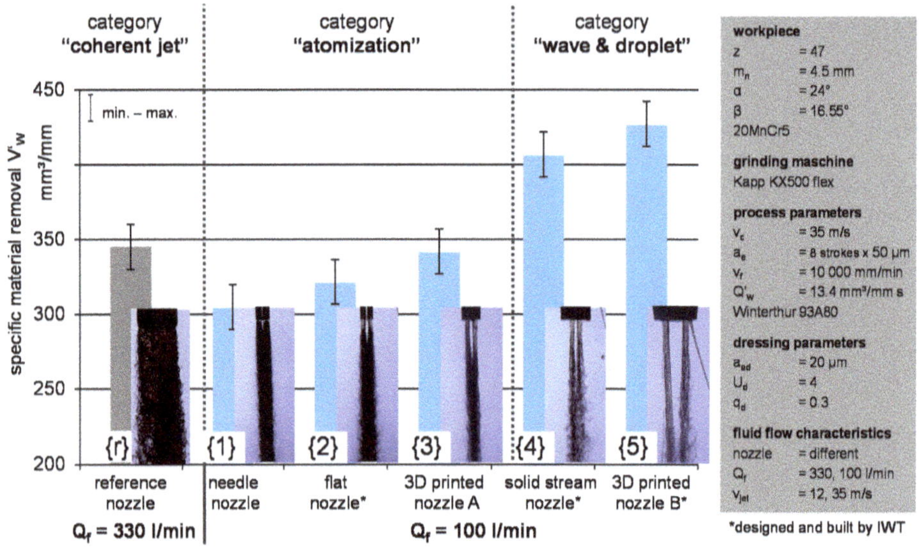

Figure 10. Grinding tests—Influence of the jet breakup.

Using the reference nozzle {r}, a specific removed material volume of $V'_W = 345 \pm 15\ mm^3/mm$ was reached. This result can be attributed to the much higher grinding fluid flowrate ($Q_f = 330\ l/min$) and the coherent jet. For the nozzles of the category "atomization" and "wave & droplet", a significant lower grinding fluid flowrate of $Q_f = 100\ l/min$ and a higher jet velocity of $v_{jet} = 35\ m/s$ was adjusted which corresponds to the cutting speed v_c. With the needle nozzle {1}, the flat nozzle {2}, and the 3D printed nozzle A {3} (category "atomization"), a specific removed material volume of $V'_W = 311–349\ mm^3/mm$ was achieved. The highest specific removed material volume of $V'_W = 415–436\ mm^3/mm$ was reached with the nozzles of the category "wave and droplet". The results of the grinding tests correlate with the coherent lengths of the jets and the calculated positions in the Ohnesorge vs. Reynolds number diagram. A longer coherent jet leads to a higher specific removed material volume. This might be due to a better wetting of the grinding wheel for a "wave and droplet" jet breakup. In conclusion, for the optimized grinding fluid supply versus the reference fluid supply, the specific removed material volume can significantly be increased (up to 26%//436 mm^3/mm {5} to 345 mm^3/mm {r}) with a simultaneous decrease of the grinding fluid flow rate (reduction of 70%//100 l/min {5} to 330 l/min {r}). For a reduced grinding fluid flow rate (100 l/min), an appropriate inner design of the nozzle alone can lead to an increase in the process performance by up to 40% (436 mm^3/mm {5} to 311 mm^3/mm {1}).

5. Conclusions and Outlook

In the fluidic investigations conducted in this study, the grinding fluid jet and the jet breakup were characterized by means of high-speed photography. The impact pressure was analyzed using a high resolution piezoelectric senor. The grinding trials showed the impact of the jet characteristics on the grinding process.

The grinding fluid jet characteristics considered here were divided into three different categories "coherent jet", "wave & droplet", and "atomization", depending on the grinding fluid nozzle. The different jet characteristics should already be taken into account when designing a grinding fluid nozzle. The jet characteristics and breakup have a significant influence on how much of the workpiece volume can be removed before thermo-mechanical damage occurs. With a higher jet quality, namely a breakup that is more "wave & droplet" than "atomization", a higher specific removed material volume can be achieved by up to 40%.

Further investigations using high-speed photography are planned to consider the interaction between the grinding fluid jet and the rotating grinding wheel for different types of jet breakup. This should deliver a better knowledge of the interaction and the efficient disposal of grinding fluid on the wheel and within the contact zone leading to a better grinding performance. In addition, the previous knowledge should be transferred to other grinding processes with higher grinding wheel circumferential speeds and complex contact conditions.

Acknowledgments: The IGF-research project 18204/N of the FVA—Forschungsvereinigung Antriebstechnik e.V.—was part of the program Industrielle Gemeinschaftsforschung und—entwicklung (IGF) and was supported by the German Federal Ministry for Economics and Technology (BMWi) via the Industrial Cooperative Research Associations (AiF). The authors express their sincere thanks to the collaborating industrial partners for the support of this research.

Author Contributions: Philip Geilert and André Wagner conceived and designed the experiments; Philip Geilert performed the experiments and analyzed the data; André Wagner and Carsten Heinzel contributed to the analysis and the interpretation of results.

Conflicts of Interest: The authors declare no conflict of interest.

References

1. Brinksmeier, E.; Heinzel, C.; Meyer, L. Coolant supply conditions and their effect on the workpiece surface layer in grinding. *Prod. Eng. Res. Dev. Ann. WGP* **2001**, *8*, 9–12.
2. Webster, J.A.; Cui, C.; Mindek, R.B., Jr. Grinding fluid application system design. *CIRP Ann.* **1995**, *44*, 333–338. [CrossRef]
3. Rowe, W. 8-Application of Fluids. In *Principles of Modern Grinding Technology*, 2nd ed.; William Andrew Publishing: Boston, MA, USA, 2014.
4. Kirsch, B. The impact of contact zone flow rate and bulk cooling on the cooling efficiency in grinding applying different nozzle designs and grinding wheel textures. *CIRP J. Manuf. Sci. Technol.* **2017**, *18*, 179–187. [CrossRef]
5. Webster, J.; Brinksmeier, E.; Heinzel, C.; Wittmann, M.; Thöns, K. Assessment of grinding fluid effectiveness in continuous-dress creep feed grinding. *CIRP Ann.* **2002**, *51*, 235–240. [CrossRef]
6. Wittmann, M.; Heinzel, C.; Brinksmeier, E. Evaluating the efficiency of coolant supply systems in grinding. *Prod. Eng. Res. Dev. Ann. WGP* **2004**, *11*, 39–42.
7. Meyer, L.; Heinzel, C.; Brinksmeier, E. Analysis and optimization of coolant supply conditions in grinding. *Prod. Eng. Res. Dev. Ann. WGP* **2005**, *12*, 27–30.
8. Cui, C. Experimental Investigation of Thermofluids in the Grinding Zone. Ph.D. Thesis, University of Connecticut, Fairfield, CT, USA, 1995.
9. Li, D. *Encyclopedia of Microfluidics and Nanofluidics*; Springer: New York, NY, USA, 2008.
10. Ohnesorge, W.V. Die bildung von tropfen an düsen und die auflösung flüssiger strahlen. *J. Appl. Math. Mech.* **1936**, *16*, 355–358. [CrossRef]
11. Oertel, H. *Prandtl—Führer Durch die Strömungslehre—Grundlagen und Phänomene*, 13th ed.; Springer Vieweg: Karlsruhe, Germany, 2012.

inventions

MDPI

Article

Experimental Analysis for the Use of Sodium Dodecyl Sulfate as a Soluble Metal Cutting Fluid for Micromachining with Electroless-Plated Micropencil Grinding Tools

Peter A. Arrabiyeh [1,*], Martin Bohley [1], Felix Ströer [2], Benjamin Kirsch [1], Jörg Seewig [2] and Jan C. Aurich [1]

[1] Institute for Manufacturing Technology and Production Systems, University of Kaiserslautern, P.O. Box 3049, 67653 Kaiserslautern, Germany; martin.bohley@mv.uni-kl.de (M.B.); benjamin.kirsch@mv.uni-kl.de (B.K.); fbk@mv.uni-kl.de (J.C.A.)

[2] Institute for Measurement and Sensor-Technology, University of Kaiserslautern, P.O. Box 3049, 67653 Kaiserslautern, Germany; stroeer@mv.uni-kl.de (F.S.); seewig@mv.uni-kl.de (J.S.)

* Correspondence: peter.arrabiyeh@mv.uni-kl.de; Tel.: +49-631-205-5483

Received: 29 September 2017; Accepted: 8 November 2017; Published: 10 November 2017

Abstract: Microgrinding with micropencil grinding tools (MPGTs) is a flexible and economic process to machine microstructures in hard and brittle materials. In macrogrinding, cooling and lubrication are done with metal cutting fluids; their application and influence is well researched. Although it can be expected that metal cutting fluids also play a decisive role in microgrinding, systematic investigations can hardly be found. A metal cutting fluid capable of wetting the machining process, containing quantities as small as 0.02% of the water-soluble fluid sodium dodecyl sulfate was tested in microgrinding experiments with MPGTs (diameter ~50 μm; abrasive grit size 2–4 μm). The workpiece material was hardened 16MnCr5.

Keywords: microgrinding; sodium dodecyl sulfate; metal cutting fluid; microstructures; micropencil grinding tools

1. Introduction

Microcomponents with functional surfaces are becoming an integral part [1] in precision industries such as biomedicine, aerospace, microelectronics and telecommunications [2]. There is a growing need for microstructured components, especially ones that manage fluids on the microscopic scale. Microfluidic components generally use a small sample volume, have a good temperature control and can downscale analytical equipment for chemical and biomedical analysis [3].

A number of microstructuring processes have been developed to fulfill the market's need for microstructured components. Processes like LIGA (lithography, electroplating, and molding), micro molding and chemical etching techniques are suitable for the mass production of these parts, but lack the flexibility for small batch production [4]. Micro-end milling [5] and microdrilling are far more suitable for small batch production lines, but are limited in the hardness of machinable materials [4].

In conventional machining, abrasive processes like grinding use superabrasives made of diamonds or cBN (cubic boron nitride) grits to machine hard and brittle materials [6]. In the past, miniaturized versions of these abrasive processes have been developed for microstructuring purposes; among them microgrinding [1]. Microgrinding has a competitive edge over other microstructuring process, since it is used as a finishing process that manufactures surfaces with optical quality while minimizing burr formation [2]. There are two kinds of microgrinding tools in micromachining [4]: thin grinding wheels called dicing blades used in the semiconductor industry to cut silicon wafers and produce open structures and micropencil grinding tools (MPGTs), used for freeform surfaces and microholes [7].

Microgrinding is a process in which the material removal takes place by stochastically distributed grits [8]. The grits have different protrusions, resulting in material removal at different chip thicknesses. The value of the chip thickness needs to surpass a minimal size to initiate material removal, otherwise only elastic and plastic deformation occur during the machining process. Most grits do not reach that minimal chip thickness, resulting in very high local temperatures due to friction [9]. High temperatures in turn can cause intense tool wear and a high thermal strain on the workpiece surface that causes material structure changes and high tensile stresses [9]. A metal cutting fluid (MCF) is needed to improve the surface quality of machined materials to transport the produced chips away from the machining zone and to increase tool life [10].

A typical flood supply of MCF is generally not suitable for micromachining, as a high flow pressure of liquids may influence the tool behavior by applying an additional force [11]. In literature, new lubrication methods and metal cutting fluids have been developed for micromachining processes. Brudek et al. compared the roughness values: the arithmetic mean roughness R_a and the mean roughness depth R_z of micromilled substrates machined with a minimum quantity method to a process where the workpiece is completely immersed in metal cutting fluid. A variety of commercially available MCFs as well as a variety of vegetable oils were used. Both lubrication methods showed similar results with the commercial MCFs having a small edge over the vegetable oils [12]. Nam et al. used nanofluids in a microdrilling process as an alternative to conventional MCFs. Nanofluids are composed by a water- or oil-based fluid containing nanoparticles made of materials like graphite, Al_2O_3, C_{60} or diamond to increase the thermal conductivity and/or decrease friction in the contact area via the ball bearing effect [13]. Pham et al. used another alternative MCF by spraying ionic liquids in a micromilling process while machining aluminum workpieces. Ionic liquids are liquid salts, consisting of an organic cation and an inorganic anion; they have a low vapor pressure, are non-flammable and have a high thermal stability. The test series showed that workpieces machined with ionic fluids reach similar cutting forces and similar roughness values to those machined with commercially available metal cutting fluids [14]. Overall, the impact of MCF in microgrinding is rarely investigated.

The machine tool used in this paper is located in a clean room and is not capsuled to its environment; a metal cutting fluid safe for both the user and the machinery is hence needed for lubrication. Sodium dodecyl sulfate (SDS) is a surfactant and detergent that has lubricating qualities. It is cheap, causes no health hazards and is even commonly used in the soap and shampoo industry [15].

In this paper, MPGTs with a diameter of ~50 μm and a cBN grit size of 2–4 μm are used to machine hardened 16MnCr5 steel using distilled water and minimal quantities of the solid surfactant SDS to lubricate the process. The results are then compared to a dry microgrinding process and one that uses distilled water only as a coolant. Besides the wear and roughness analysis, the paper offers a tool characterization pre- and post-machining, a structure characterization as well as a force analysis.

2. Materials and Methods

2.1. Micropencil Grinding Tools

The substrates used for the MPGTs contain a tungsten carbide content of 92%, a cobalt content of 8% with a grain size of 0.2 μm. The shaft of the substrates has a diameter of 3.175 mm, a bending strength of 4800 N/mm^2 and a Vickers hardness of 1920 ± 50 HV30 (ISO 3878) [16]. Figure 1 depicts the geometry of a substrate and the two main steps in the manufacturing of MPGTs. A 40° cone is machined on a conventional tool grinding machine onto the substrate to decrease the material removal in the following precision grinding steps. Using a thin grinding wheel, the cylindrical tip of the substrate is machined to have a diameter of 44 ± 2 μm at a length of 140 μm. For this paper, a grit size of 2–4 μm is used. The tool tip diameter must be readjusted when using different grit sizes with different coating thicknesses to allow the coated tool to reach a diameter of ~50 μm.

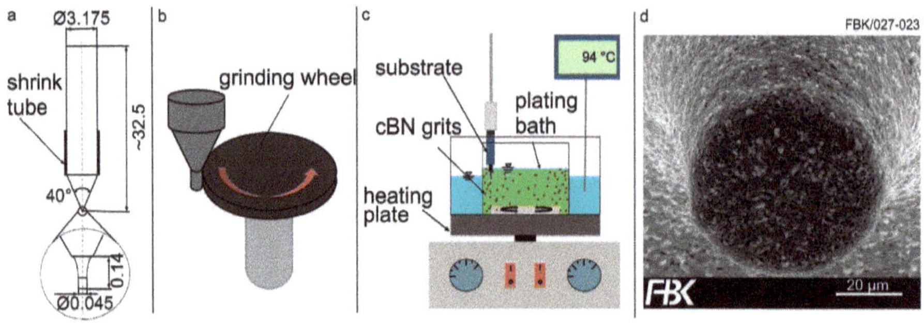

Figure 1. Manufacturing process for micropencil grinding tools (MPGTs): (**a**) geometry of machined substrate; (**b**) microgrinding process for MPGT substrates; (**c**) electroless plating process for MPGT and (**d**) finished MPGT with 2–4 µm grit size.

Following the machining process, the substrate is degreased in an alkaline degreasing solution, which is then neutralized in a hydrochloric acid solution. Then a thin nickel layer is electroplated onto the substrate (Figure 2) to provide an active, chemically affine nickel surface. Shrink tubes are applied to the substrates as a resist to limit the nickel coating to a defined area [17]. Finally, the electroless-plating process is performed.

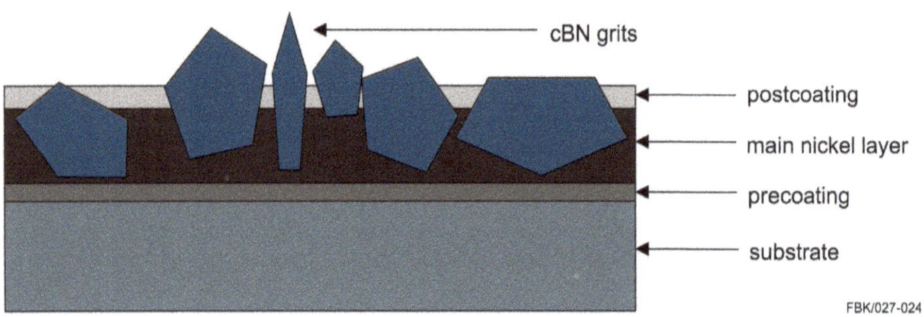

Figure 2. Abrasive layer. cBN: cubic boron nitride.

Electroless plating is a process that is based on the principle of ionic reduction. In the solution, a metal salt, in this case nickel sulfate, provides the solution with Ni^{2+} free nickel ions. A reducing agent like sodium hypophosphite can provide the nickel ions with the two missing electrons, slowly forming a phosphorous nickel coating onto an active surface; the components of the plating solution are listed in Table 1. The process is suited to manufacture small quantities of MPGTs, with flexible, custom design choices in regards to its form, diameter, coating thickness, grit size, grit concentration and grit protrusion [17].

Using the ingredients listed in Table 1, a quantitative energy dispersive X-ray (EDX) analysis shows that a phosphorous content of 6.01% ± 0.55% can be achieved. A phosphorus content of less than 7% results in a face-centered cubic crystal structure, while an amorphous structure is produced at higher phosphorus contents [18]. A low phosphorous content generally produces a harder nickel layer [19].

Table 1. Electroless-plating solution composition [20].

Component	Concentration in g/L
Nickel sulfate ($NiSO_4 \cdot 6H_2O$)	30
Sodium hypophosphite (NaH_2PO_2)	20
Sodium acetate ($C_2H_3NaO_2$)	20
Thiourea (CH_4N_2S)	0.0004
Hydrochloric acid (HCl)	Adapted to a pH value of 5.2–5.4
cBN grits	4

The abrasive grits are whirled up in the coating solution via a magnetic stirrer. The main coating time for a monolayered MPGT is 150 s for a grit size of 2–4 µm. After the main coating time, the magnetic stirrer stops and the grits fall to the bottom of the beaker, allowing to embed the grits on the MPGT with an additional nickel layer for 90 s (see Figure 2). The final product for a single layered MPGT can be seen in Figure 1d. Using scanning electron microscopy (SEM) images and an image processing software, a quantitative analysis was conducted to determine the grit concentration on the tool. A grit concentration of 35% ± 7% was found.

2.2. Experimental Setup

The results presented in the following chapters were produced on a high precision three-axis machine tool (Figure 3) mounted on top of a vibration isolated granite plate. The tool spindle is mounted vertically onto the z-axis on a cross-roller bearing stage. Rotation speeds in the range of 5000–54,000 rpm can be achieved. The X–Y table is guided by air bearings and can move with a positioning accuracy of <1 µm [5]. A Kistler 3-component dynamometer (9119AA1) dynamometer for measuring cutting forces is mounted on top of the X–Y table; the workpieces are clamped onto the dynamometer.

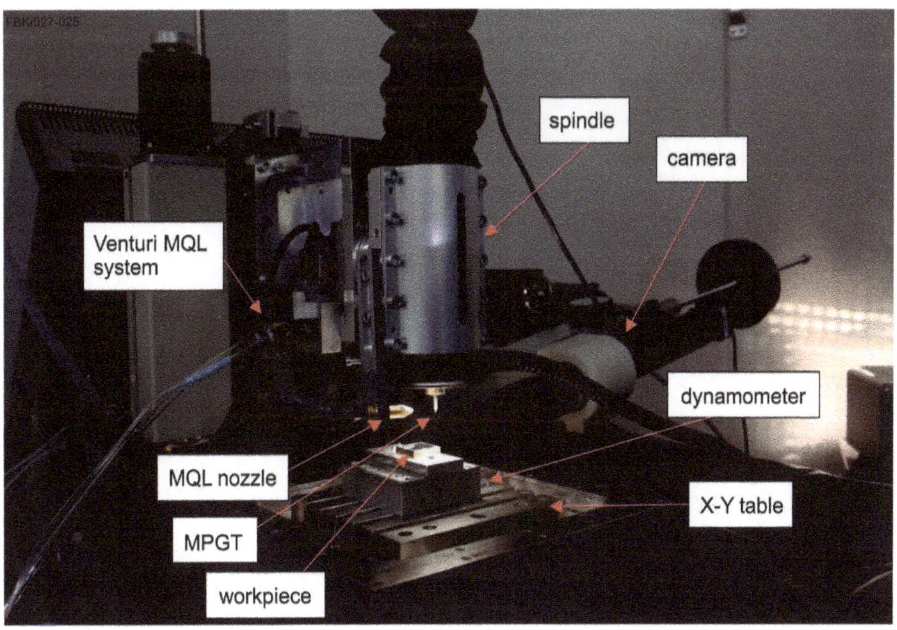

Figure 3. Machine tool for microgrinding and milling. MQL: minimum quantity lubrication.

A Venturi minimum quantity lubrication (MQL) system is used to spray the machining process with an air/liquid mixture. By narrowing the cross-section at the nozzle head, a pressure difference is created through which the liquid is suctioned. The air acts as a transport medium for the liquid. Flow rates of <100 mL/h are defined as MQL in macro machining. However, considering the small sizes of tools and structures and the comparably low material removal rates in microgrinding, 100 mL/h can be defined as flood cooling in microgrinding.

Images of the tools and their respective structures were captured using a scanning electron microscope (SEM). The structures were analyzed using a confocal microscope (Nanofocus µsurf) with a 60× magnification lens and a numerical aperture (NA) of 0.9.

2.3. Experimental Procedure

To test the influence of SDS as a metal cutting fluid in the microgrinding process, MPGTs with diameters ~50 µm were used to machine 500 µm long grooves into hardened 16MnCr5 (SAE5115; 665 HV30 ± 15 HV30 (according to ISO 6507 [21])). The workpiece was face-machined with a larger pencil grinding tool (diameter = 3.175 mm) to compensate for assembly-related influences and to gain a flat surface. To test the effect of the soluble lubricant, a small amount of 0.2 g/L was added to a distilled water medium and was used for the experiments. For comparison, the microgrinding process was also conducted dry and with distilled water as metal cutting fluid. Both the SDS and distilled water experiments were conducted with a volume flow rate of 60 ± 10 mL/h and a positive air pressure of 0.65 bar. Both form a fluid film around the tool during the process (Figure 4b); experiments showed that if the fluid film is interrupted, immediate damage to the abrasive layer occurs.

Based on preliminary studies; a rotation speed of 30,000 rpm (cutting speed of 4.71 m/min) was applied. Feed rates of 0.05 mm/min and 0.1 mm/min were used at a depth of cut of 5 µm; Figure 4a shows the parameter combinations studied in this paper with each combination being repeated three times. The tools were maneuvered to the starting position optically using the camera. The MPGT is used to scratch the surface of the workpiece to determine the zero position between tool and workpiece. This results into a positioning accuracy of ±50 µm and hence in an according deviation of the groove length. The tool rotates in clockwise direction, while the workpiece was given a feed rate towards the tool.

a

case	feed rate in mm/min	rotational speed in rpm	metal cutting fluid
case 1	0.05	30,000	dry
case 2	0.1	30,000	dry
case 3	0.05	30,000	water
case 4	0.1	30,000	water
case 5	0.05	30,000	SDS
case 6	0.1	30,000	SDS

b

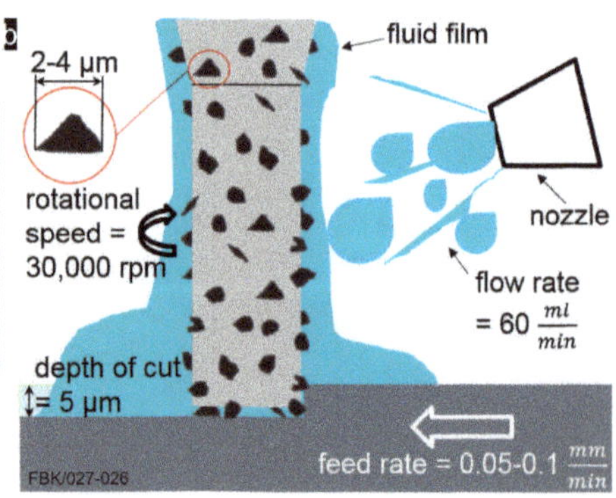

Figure 4. (**a**) Microgrinding test series; (**b**) microgrinding process.

3. Results

3.1. Tool Wear

Figure 5 visualizes the tool wear for four of the parameter combinations (cases) listed in Figure 4a; the results from parameter combination 2 and 4 were left out of the figure, because they do not differ from cases 1 and 3. The tools used in the dry machining process lose their abrasive layer on the face side of the tool upon entry (Figure 5a). The abrasive layer breaks off the MPGT and rips part of the layer on the circumference of the tool as well (Figure 5a). It is assumed that high temperatures occur on the face side of the MPGT due to friction. Tungsten carbide with an 8% cobalt content has a linear thermal expansion coefficient of 5×10^{-6}–5.2×10^{-6} K^{-1} which is much smaller than that of nickel which lies in a range of 12×10^{-6}–13.5×10^{-6} K^{-1} [22]. Thus, the abrasive layer expands much more with rising temperature than the substrate, causing the abrasive layer to loosen up until the process forces eventually result into failure of the abrasive layer.

MPGTs cooled with water showed a slight improvement over the ones used in dry machining, despite losing the abrasive layer. The tools had a longer tool life than the ones used in dry machining as can be seen from the abrasive layer breakoff point presented in Figure 5b. The abrasive layer stays in tact in both cases that use SDS for lubrication (Figure 5c,d); except for one tool used in case six that broke off 10 μm before finishing its groove. Tool wear is slightly higher for case 6 due to the increase in feed rate.

An energy-dispersive X-ray analysis (EDX) was performed to determine the iron adhesion on the face side of the abrasive layer and the nickel adhesion at the bottom of the structure. While the analysis showed no signs of nickel adhesion in the structure, an iron concentration of 28.65% ± 5.1% for case 5 and 50.05% ± 0.78% for case 6 was identified. The results are qualitative and require further investigation; however, a direct relationship between feed rate and material adhesion could be determined. According to Klocke, adhesions that fill the chip space, increase friction, and therefore the process temperature and the process forces. Material adhesions increase the wear of the machining tool by breaking out single grits or even entire grit populations [9].

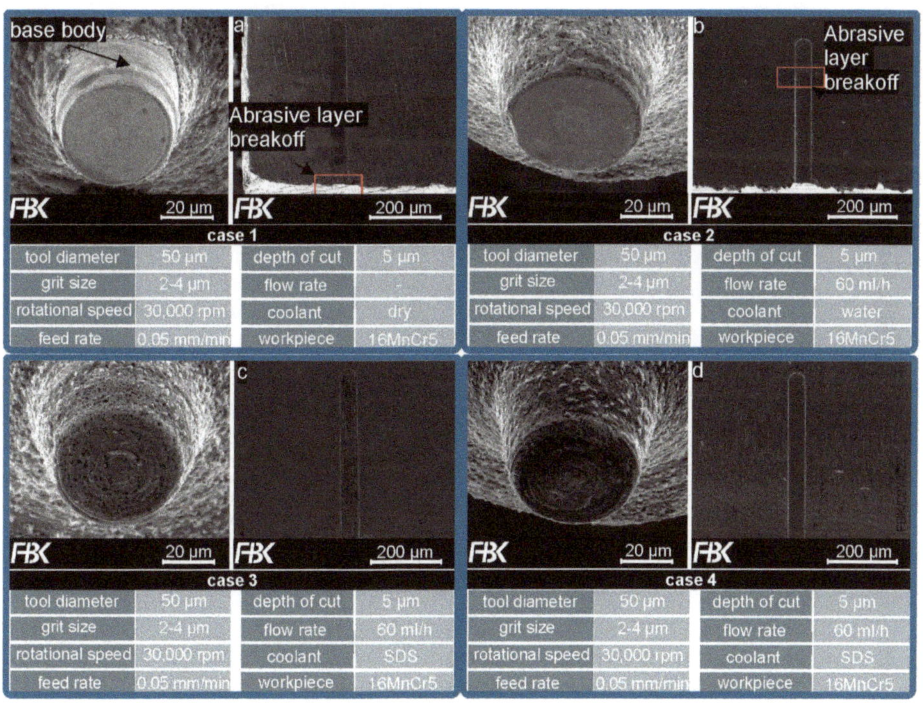

Figure 5. MPGTs post-microgrinding: (**a**) case 1; (**b**) case 3; (**c**) case 5 and (**d**) case 6.

3.2. Structure Analysis

Topographies of the structures machined with all six parameter combinations were measured using a confocal microscope (Nanofocus µsurf) with a 60× magnification lens and a numerical aperture of NA = 0.9. Multiple measurements were combined using the stitching algorithms integrated in the measurement devices software. Missing data points, resulting from the finite numerical aperture or artefacts on the surfaces, were interpolated using linear interpolation. A first order plane levelling minimizing the sum of the squared distances was applied to the areal measurement data. Profiles located in the center of the groove with a length of 400 µm were extracted manually from the areal data. Calculation of 2D roughness parameters included limitation of the bandwidth using the Gaussian-filter [23,24] (l_c = 80 µm; l_s = 0 µm).

To estimate the surface quality, the mean roughness depth R_z and the arithmetical mean roughness value R_a [25], were calculated according to DIN EN ISO 4288. According to the ISO standard, five consecutive 80 µm long segments (from a 400 µm long section) were used to calculate the roughness R_{ai} with i = 1, . . . , 5 of a structure [26], with the first segment starting at the entry point of each groove, in order to monitor the tool right from the start.

Figure 6a,b display the mean values of R_a and R_z for all six parameter combinations with their respective standard deviations. A few trends are visible in the diagrams; the first being that grooves machined with lubrication (Cases 5 and 6), resulted in the smallest roughness values, while grooves machined dry exhibited overall higher values. The standard deviation of the roughness parameters is also much smaller for grooves machined with the SDS mixture, which is synonymous for a more stable process, since less material adhesion on the slot bottom appeared. Another visible trend is that grooves machined with a feed rate of 0.1 mm/min feature smaller roughness values when compare to those machined with a 0.05 mm/min feed rate which is a surprising trend considering higher uncut chip thicknesses at higher feed rates. The lower roughness at higher feeds could be led back to the higher tool wear and higher material adhesion at larger feed rates.

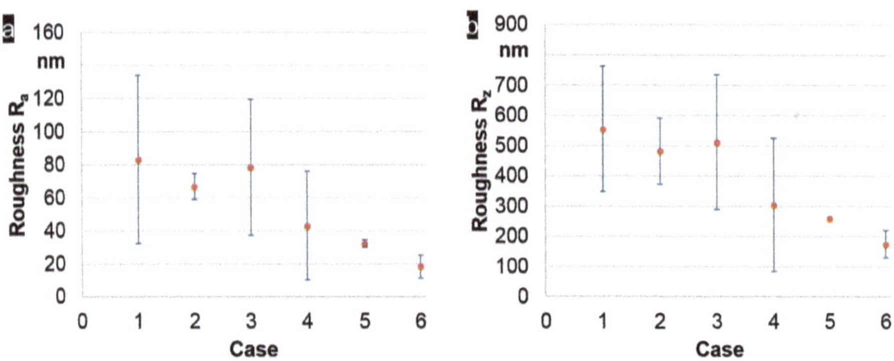

case	feed rate in mm/min	rotational speed in rpm	metal cutting fluid
case 1	0.05	30,000	dry
case 2	0.1	30,000	dry
case 3	0.05	30,000	water
case 4	0.1	30,000	water
case 5	0.05	30,000	SDS
case 6	0.1	30,000	SDS

FBK/027-028

Figure 6. (**a**) Roughness values R_a and (**b**) Roughness values R_z.

The sample structure presented in Figure 7c is one of the structures from case 6. It is chosen to present some of the more prominent tool-specific and structural characteristics. The first being the obvious breakoff at the entry point—this indicates a loss in abrasive grits. Judging from Figure 7a, many grits at the center of the tool had a rather high protrusion; some of these grits broke off upon entry due to a lack of grit retention forces at small high grit protrusion and at the same time high uncut chip thickness, resulting in high loads on those grits. A small 9 µm broad smaller groove in the middle of the structure (Figure 7c) suggests that the grits at the center of the tool had a higher protrusion. Figure 7b shows the tool after the machining process and a small circular marking off the pivot of the tool, which coincidently measures to 9 µm. This pivot is located 2.5 µm off the center of the tool, hence a reclamping error occurred, producing a step-like structure at the right side of the groove (Figure 7c). This difference in cutting depths can be led back to a difference in height for parts of the abrasive layer; this difference in height is marked in Figure 7b but can also be seen in Figure 7a (less grit protrusion). One final, more common characteristic is the material adhesion to the bottom of the groove, on the up grinding side.

The manifestation of these characteristics is unique to each individual MPGT. Some of them can be prevented by sorting out tools that have a higher variation in grit protrusion; however, due to constant wear, the grits are exposed to during the machining process—it is impossible to prohibit them completely.

The groove analyzed in in Figure 7c was machined with SDS and a feed rate of 0.1 mm/min. Different from other structures machined with SDS, the structure has a large material adhesion at the bottom surface; a characteristic more commonly observed with those machined dry. The structure in Figure 7c shows almost no signs of burrs and only few, small chippings, while cases 1–4 show much larger and frequent burr and chipping formations because no material cutting happens after the loss of an abrasive layer. An example of a dry machined structure is shown in Figure 8.

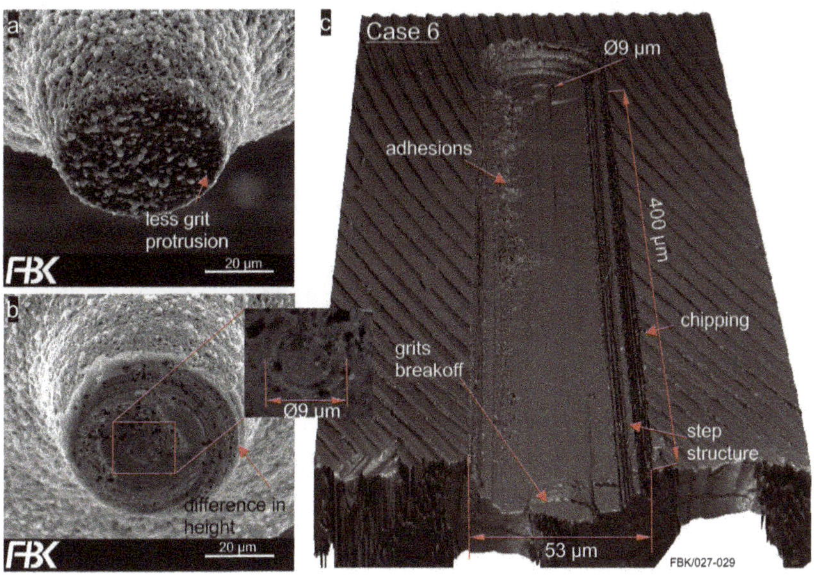

Figure 7. (**a**,**b**) Tool that machined the groove sample, before and post machining and (**c**) sample of a structure machined with sodium dodecyl sulfate (SDS) and a feed rate of 0.1 mm/min, measured by confocal microscopy.

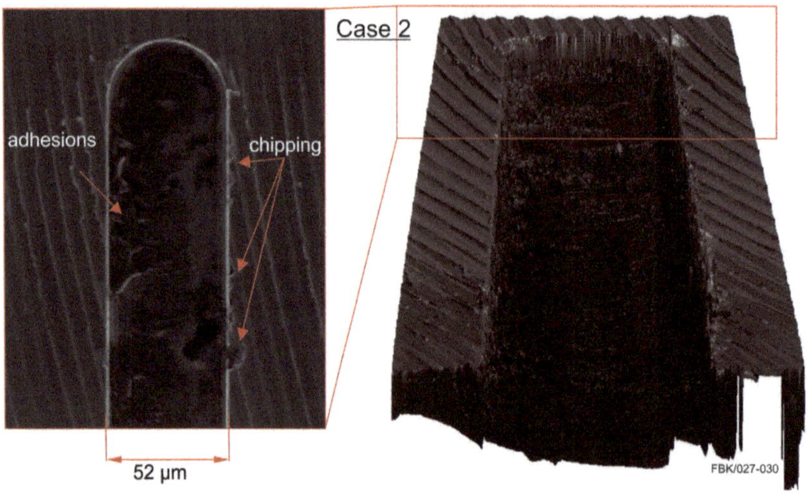

Figure 8. Sample of a dry machined structure with a feed rate of 0.1 mm/min, measured by confocal microscopy. A segment is shown with scanning electron microscopy (SEM).

3.3. Force Measurements

The process forces were measured with a Kistler 3-component dynamometer (9119AA1) during machining. The sampling rate was 10 kHz for all grinding parameters.

After recording, the data was evaluated with National Instruments Software DIAdem. With the help of a fast Fourier transformation, the actual spindle speed was determined. With this information a bandpass filter was applied to extract and consider the frequencies in the signal corresponding to the spindle speed ±20 Hz only. Thus, not the actual process forces were used for this research but the dynamic process characteristics evaluated and compared. The reason for this configuration is the use of metal cutting fluid. The impact of the fluid application on the workpiece was higher than the impact of the cutting process itself.

Figure 9 shows one of the grooves machined for case 4 to demonstrate the correlation between forces and depth of cut. In Section 1, where the tool is starting to cut the material, the force rates are rising until the complete tool diameter is in contact. This area is followed by Section 2, where the cutting conditions seem to be unstable. The cutting forces as well as the feed forces start to rise and show sharp peaks. At a feed travel of about 150 µm, the forces reach their maximum and some abrasive grits broke out of the layer or parts of the layer itself broke out, which can be seen in the confocal image, showing a difference in the resulting geometry of the slot bottom. Until a feed travel of 250 µm (Section 3), the surface as well as the force amplitudes show an unstable behavior. The depth of cut is changing in two steps. Until the end of the slot at 500 µm, the force levels as well as the slot bottom geometry are at a stable cutting regime. At the end of Section 4, the force levels are falling abruptly after the feed of the machine tool is being stopped.

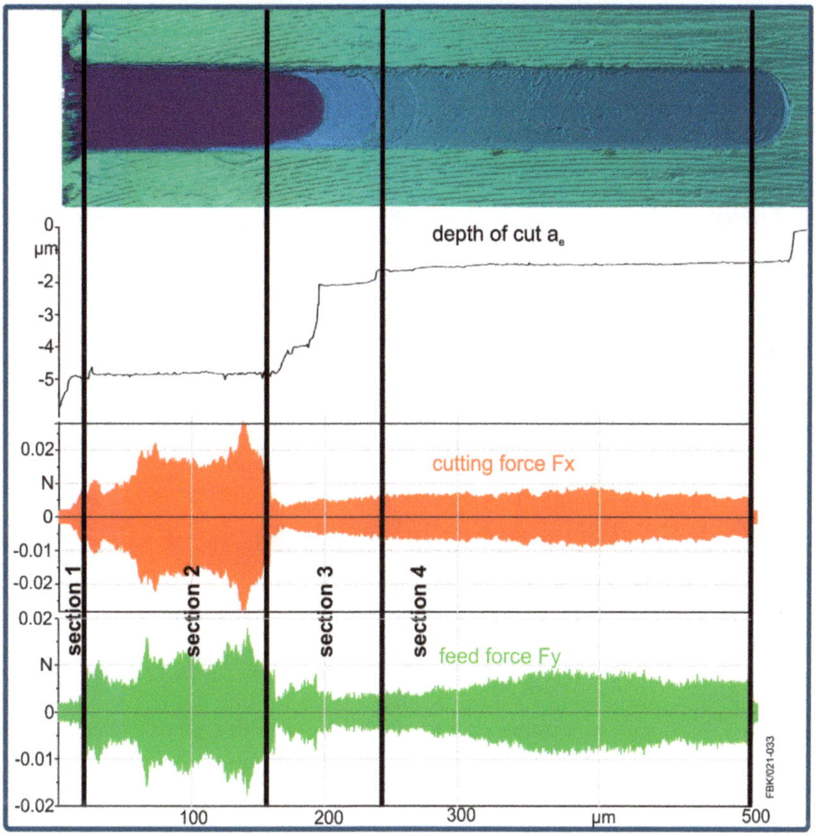

Figure 9. Comparison of slot bottom geometry and the equivalent force levels.

4. Conclusions and Outlook

Microgrinding is a process in which a very high amount of rubbing and ploughing occurs. This can result in high temperatures, high wear of the tools and ultimately a low quality of the machined structures. This paper presents a systematic investigation on the influence of the application of metal cutting fluids on tool wear and structure quality.

The application of a new metal cutting fluid (MCF), consisting of small quantities (0.02% in distilled water) of the solid lubricant sodium dodecyl sulfate (SDS) was tested when microgrinding. In this test series, microgrinding experiments with monolayered electroless plated micropencil grinding tools (diameter ~50 µm and grit size 2–4 µm) were conducted to analyze the effect of the new metal cutting fluid. The results were compared to pure distilled water and dry machining experiments at a rotational speed of 30,000 rpm and the two-feed rate variations of 0.05 mm/min and 0.1 mm/min.

The results showed that the micropencil grinding tools (MPGTs) used in dry machining and with distilled water were unable to complete a 500-µm long groove without losing part of their abrasive layer. In contrast, the grooves machined with SDS were completed with the tool intact.

The structures were analyzed using a confocal microscope. Both the roughness values and the structure characteristics were examined. Highest roughness values were measured for the dry grinding cases and lowest roughness values for cases machined with SDS. Besides the lower roughness, a more stable process was achieved with the application of SDS. This was manifested by small standard deviations of the roughness values. The application of SDS also resulted into less burr formation and chipping, as well as less adhesions on the bottom surface.

In conclusion, the results revealed that the process stability and the quality of the machined structures is highly influenced by the application of metal cutting fluids. Future work will hence deal with the investigation of the flow rate and the SDS concentration. In addition, tool and machining parameter case studies are required to explore the possibilities and limitations of increasing both the tool life and the productivity of the process. A further analysis of structure and tool characteristics will be conducted to get a better understanding of the tool wear characteristics and their influence on the structures.

Acknowledgments: This research was funded by the German Research Foundation (DFG) within the Collaborative Research Center 926 "Microscale Morphology of Component Surfaces"—through the subproject B09 "Geometrical Structuring of Component Surfaces by Microgrinding".

Author Contributions: Peter A. Arrabiyeh has conceived, designed, performed and documented the microgrinding and electroless plating experiments. Peter A. Arrabiyeh, Martin Bohley and Felix Ströer conducted the structural analysis of machined microstructures. Martin Bohley analyzed the data collected from the dynamometer. Peter A. Arrabiyeh conducted SEM and EDX analysis. Benjamin Kirsch supervised the present study and helped to discuss and analyze the results. Jan C. Aurich initiated the study. All authors were highly involved in writing the paper.

Conflicts of Interest: The authors declare no conflicts of interest.

References

1. Feng, J. Microgrinding of Ceramic Materials. Ph.D. Thesis, Michigan Technological University, Houghton, MI, USA, 2010.
2. Pratap, A.; Patra, K.; Dyakonov, A.A. Manufacturing miniature products by micro grinding: A review. *Procedia Eng.* **2016**, *150*, 969–974. [CrossRef]
3. Wensink, H. Fabrication of Microstructures by Powder Blasting. Ph.D. Thesis, University of Twente, Enschede, The Netherlands, 2002.
4. Engmann, J. *Galvanisch Gebundene Mikroschleifstifte. Entwicklung, Herstellung und Einsatz, Als Ms. Gedr;* Produktionstechnische Berichte aus dem FBK 01/2011; Technischen Universität Kaiserslautern: Kaiserslautern, Germany, 2011.
5. Reichenbach, I.G.; Aurich, J.C. Untersuchung der oberflächengüte beim mikrofräsen—einfluss von prozessparametern und einstellwinkel der nebenschneide bei 48 μm mikroschaftwerkzeugen. *Wt Werkstattstechnik Online* **2013**, *103/11–12*, 847–852.
6. Brinksmeier, E.; Mutlugünes, Y.; Klocke, F.; Aurich, J.C.; Shore, P.; Ohmori, H. Ultra precision grinding. *CIRP Ann. Manuf. Technol.* **2010**, *59*, 652–671. [CrossRef]
7. Hoffmeister, H.-W.; Hlavac, M. Schleifen von Mikrostrukturen. In *Tagungsband des 10. Feinbearbeitungskolloqiums in Braunschweig;* Vulkan-Verlag: Essen, Germany, 2002; pp. 7.1–7.24.
8. Setti, D.; Sinha, M.K.; Ghosh, S.; Rao, P.V. Performance evaluation of Ti–6Al–4V grinding using chip formation and coefficient of friction under the influence of nanofluids. *Int. J. Mach. Tools Manuf.* **2015**, *88*, 237–248. [CrossRef]
9. Klocke, F. *Manufacturing Processes 2: Grinding, Honing, Lapping;* Springer: Berlin, Germany, 2009.
10. Brinksmeier, E.; Heinzel, C.; Wittmann, M. Friction, cooling and lubrication in grinding. *CIRP Ann. Manuf. Technol.* **1999**, *48*, 581–598. [CrossRef]
11. Dornfeld, D.; Min, S.; Takeuchi, Y. Recent advances in mechanical micro-machinability of copper 101 using tungsten carbide micro-end mills. *Int. J. Mach. Tools Manuf.* **2006**, *55*, 745–768.
12. Brudek, G. *Beiträge zur Prozessanalyse in der Mikrozerspanung. Insbesondere für das Mikrofräsen;* Berichte aus dem Institut für Konstruktions- und Fertigungstechnik Bd. 8; Shaker: Aachen, Germany, 2007.
13. Nam, J.S.; Kim, D.H.; Chung, H.; Lee, S.W. Optimization of environmentally benign micro-drilling process with nanofluid minimum quantity lubrication using response surface methodology and genetic algorithm. *J. Clean. Prod.* **2015**, *102*, 428–436. [CrossRef]
14. Pham, M.Q.; Yoon, H.S.; Khare, V.; Ahn, S.H. Evaluation of ionic liquids as lubricants in micro milling-process capability and sustainability. *J. Clean. Prod.* **2014**, *76*, 167–173. [CrossRef]

15. Smulders, E.; Rybinski, W.-V.; Sung, E.; Rähse, W.; Steber, J.; Wiebel, F. *Nordskog: Ullmann's Encyclopedia of Industrial Chemistry*; Wiley-VCH Verlag GmbH & Co. KGaA: Weinheim, Germany, 2000.
16. *DIN EN ISO 3878, Hardmetals, Vickers Hardness Test, Identical with ISO 3878:1983*; International Organization for Standardization: Geneva, Switzerland, 1991.
17. Arrabiyeh, P.A.; Kirsch, B.; Aurich, J.C. Development of micro pencil grinding tools via an electroless plating process. *J. Micro Nano Manuf.* **2017**, *5*, 011002. [CrossRef]
18. Lin, K.L.; Hwang, J.W. Effect of thiourea and lead acetate on the deposition of electroless nickel. *Mater. Chem. Phys.* **2002**, *76*, 204–211. [CrossRef]
19. Mallory, G.O.; Hajdu, J.B. *Electroless Plating: Fundamentals and Applications*; American Electroplaters & Surface Finishers Society: Orlando, FL, USA, 1990.
20. Kirsch, B.; Bohley, M.; Arrabiyeh, P.A.; Aurich, J.C. Application of ultra-small micro grinding and micro milling tools: Possibilities and limitations. *Micromachines* **2017**, *8*, 261. [CrossRef]
21. *DIN EN ISO 6507-1:2006-03 Metallic Materials—Vickers Hardness Test—Part 1: Test Method*; International Organization for Standardization: Geneva, Switzerland, 2006.
22. Granta Design Ltd. *CES EduPack Software*; Granta Design Ltd.: Cambridge, UK, 2012.
23. *DIN EN ISO 16610-21:2013-06 Geometrical Product Specifications (GPS)—Filtration—Part 21: Linear Profile Filters: Gaussian Filters*; International Organization for Standardization: Geneva, Switzerland, 2012.
24. *DIN EN ISO 11562:1998-09 Geometrical Product Specifications (GPS)—Surface Texture: Profile Method—Metrological Characteristics of Phase Correct Filters*; International Organization for Standardization: Geneva, Switzerland, 1997.
25. *DIN EN ISO 4287:2010-07 Geometrical Product Specifications (GPS)—Surface Texture: Profile Method—Terms, Definitions and Surface Texture Parameters*; International Organization for Standardization: Geneva, Switzerland, 2009.
26. *DIN EN ISO 4288:1998-08 Geometrical Product Specifications (GPS)—Surface Texture: Profile Method—Rules and Procedures for the Assessment of Surface Texture*; International Organization for Standardization: Geneva, Switzerland, 1997.

inventions MDPI

Article

Model Development for Optimum Setup Conditions that Satisfy Three Stability Criteria of Centerless Grinding Systems

Fukuo Hashimoto

Advanced Finishing Technology Ltd., Akron, OH 44319, USA; fukuohashimoto@gmail.com

Received: 5 September 2017; Accepted: 16 September 2017; Published: 21 September 2017

Abstract: The centerless grinding process demonstrates superior grinding accuracy with extremely high productivity, but only if the setup conditions are properly set up. Otherwise, various unfavorable phenomena manifest during the grinding processes and become serious obstacles to achieving that high quality and productivity. These phenomena are associated with the fundamental stabilities of the centerless grinding system, so it is essential to keep the system stable by setting up the appropriate grinding conditions. This paper describes the development of a model for finding the setup conditions that simultaneously satisfy the three stability criteria of centerless grinding systems: (1) work rotation stability for safe operations; (2) geometrical rounding stability for better roundness; and (3) dynamic system stability for chatter-free grinding. The objective of the model development is to produce combinations of optimal setup conditions as the outputs of the model, and to rank the priority of the outputs using PI (performance index) functions based on the process aims (productivity or accuracy). The paper demonstrates that the developed model, named Opt-Setup Master, can generate the optimum setup conditions to ensure safe operations, better roundness and chatter-free grinding. It provides practical setup conditions as well as scientific parameters and fundamental grinding parameters. Finally, the paper verifies that the Opt-Setup Master provides the setup conditions that simultaneously satisfy all three stability criteria of the centerless grinding system.

Keywords: grinding; centerless grinding; process optimization; safe operation; quality; productivity

1. Introduction

The centerless grinding method has been extensively applied for the production of cylindrical components such as rings, rollers, and pins. It is estimated that a single car has more than 2000 parts finished by centerless grinding processes. The centerless grinding process demonstrates extremely high productivity with very high grinding accuracy in OD size, roundness and surface integrity. However, its superior performance compared to other grinding methods can be achieved only if the grinding conditions are properly set up; otherwise, various unfavorable phenomena, such as slippages in work rotation, deformed roundness and chatter vibrations, appear during the grinding process and lead to deterioration in grinding performance [1].

These huge advantages and disadvantages come from the unique work-holding features of the centerless grinding system: (1) a loose hold on the workpiece without any mechanical constraints; (2) the work friction brake/drive mechanism of the work rotation; and (3) a self-centering mechanism called "regenerative centering". The grinding process is very sensitive to these unique centerless setup conditions, so it is essential to secure the grinding system's stability by setting it up appropriately. This requires controlling the three fundamental stability issues caused by the work-holding features of centerless grinding. These are: (1) work rotation stability; (2) geometrical rounding stability; and (3) dynamic system stability.

Work rotation stability is related to the work friction brake/drive mechanism of the centerless grinding system. The regulating wheel is in rolling-sliding contact with the workpiece, and provides the friction force to the workpiece that drives or brakes the work rotation. In this unique mechanism, the work rotates with almost the same peripheral velocity of the regulating wheel during the stable grinding process, in which the torque created by the grinding force balances with the torque from the friction forces acting on the regulating wheel and the blade top surface. However, under heavy grinding with excessive grinding force, control over the work rotation speed is lost due to the broken torque equilibrium, and it increases toward the grinding wheel speed. This phenomenon, called "spinners", can cause dangerous accidents and should be avoided in order to maintain safe operations. The author is a pioneer of the study of work rotation stability and has shown that there exists an absolute safe zone where spinners do not develop [2]. The setup guidelines for safe operations are well established in the literature, and the means of satisfying the work rotation stability criterion have been demonstrated [3].

Geometrical rounding stability is related to the work-holding conditions and regenerative centering effects. Although centerless grinding technology has been around for 100 years since the method was patented by L.R. Heim in 1917 [1], a great deal of effort was exerted by early research pioneers to understand its rounding mechanism, and significant papers have been published [4–9]. The theory of the rounding mechanism has been well established, the setup guidelines for achieving better roundness have been described, and the means of satisfying the stability criteria have been clarified. The stability criteria assume that the grinding system consists of solid bodies and is dynamically stable. Under certain work-holding conditions, a specific number of lobes on the roundness of the workpiece appear or cannot be removed. It is crucial to minimize roundness errors by selecting the proper setup conditions.

Dynamic system stability is related to the work-regenerative chatter vibration caused by the instability of the centerless grinding system, including the machine dynamics. The chatter vibration in centerless grinding is very severe and builds up very fast. In general, the amplitude growth rate is 10 to 100 times greater than that of center-type grinding processes, and is caused by the wheel-regenerative chatter vibration. Significant investigations have been carried out by many researchers [10–13] to understand dynamic system stability and suppress the chatter vibration. The system stability criterion has been well established, and the setup condition guidelines for chatter-free grinding are available in the literature.

As mentioned above, the setup guidelines for satisfying each stable criterion have been established. However, the setup operations of centerless grinding still rely on experimental skill and the trial-and-error method. Even though each stable criterion can be satisfied individually by carefully choosing the setup conditions, it is almost impossible to simultaneously satisfy all three of the centerless grinding system's stability criteria. Therefore, a special analytical tool for finding the optimum combination of setup conditions is greatly needed [14].

The objective of this paper is to describe the development of an analytical model capable of finding the optimum combination of setup conditions that satisfies all three stability criteria at the same time. This paper describes the structure of the developed model, which consists of the input-information session, the data bank that stores all the parameters required for the model calculations, the PI (performance index) functions for assessing the setup conditions based on the process aim (productivity or accuracy), and the output-information session.

Further, this paper explains the algorithm of the model and shows how to find the setup conditions that meet the three stability criteria simultaneously. The developed model, named Opt-Setup Master, is verified through case studies in which workpieces with various sizes are ground with different grinding machines. Finally, the Opt-Setup Master demonstrates its capability to generate the optimum setup conditions that satisfy all three stability criteria, and to provide the grinding conditions that will provide safe grinding operations and chatter-free grinding with improved grinding accuracy.

2. Basic Setup Conditions in Centerless Grinding

The basic setup parameters in centerless grinding are the blade angle θ, the center height angle γ and the work rotational speed n_w, as shown in Figure 1.

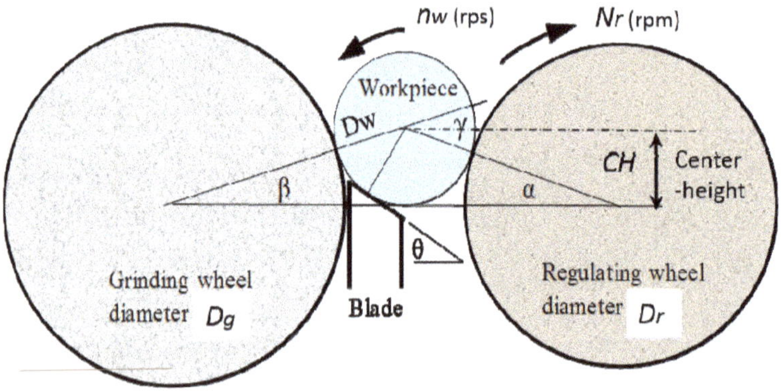

Figure 1. Setup conditions in centerless grinding.

The set (θ, γ, n_w) of these parameters is called the "setup condition" in this paper, and it significantly affects centerless grinding performance. In practice, the work center height *CH* (instead of γ) and the *RW* (regulating wheel) rotation speed N_r (rpm) are used because these parameters can be directly set up on the machine. The center height *CH* (mm) has the following relationship with the center height angle γ (°) when angles α and β are small.

$$\gamma = \alpha + \beta \tag{1}$$

$$\alpha = \sin^{-1}\frac{2CH}{(D_r + D_w)} \cong \frac{2CH}{(D_r + D_w)} \tag{2}$$

$$\beta = \sin^{-1}\frac{2CH}{(D_g + D_w)} \cong \frac{2CH}{(D_g + D_w)} \tag{3}$$

$$CH\ (\text{mm}) = \frac{3.14(D_g + D_w)(D_r + D_w)}{360(D_g + D_r + 2D_w)}\gamma(^\circ) \tag{4}$$

The work rotation speed n_w is controlled by the *RW* friction drive/brake mechanism. Figure 2 shows test results of normal grinding force *Fn*, the friction coefficient μr and the rolling-sliding velocity between *RW* and the workpiece during an infeed centerless grinding process [15]. In steady state grinding, the sliding velocity ΔV, defined as $(V_w - V_r)$, is about +0.008 m/s, and the slippage ratio $\Delta V/V_r$ is about 2%, where V_w and V_r are the work and *RW* peripheral velocities, respectively. Since the sliding velocity is very small, the work rotation speed n_w can be represented by:

$$n_w(\text{rps}) \cong \frac{D_r N_r(\text{rpm})}{60 D_w} \tag{5}$$

In the model development, the scientific parameters (θ, γ, n_w) are used for the analysis of the optimum setup condition, and the practical parameters (θ, *CH*, N_r) are the outputs of the model.

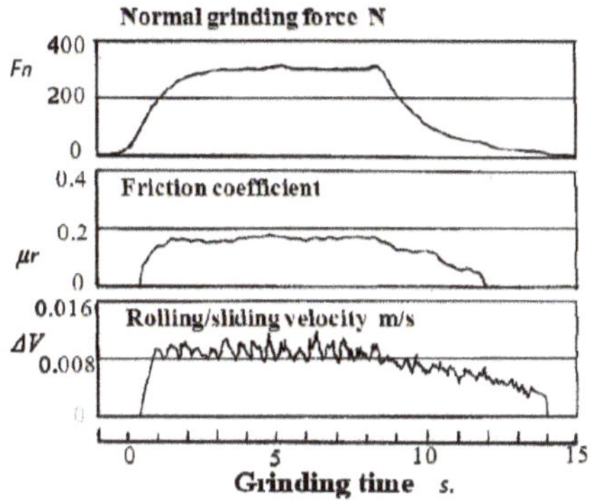

Figure 2. Infeed centerless grinding process. D_r = 255 mm, D_w = 30 mm, N_r = 30 rpm, V_w = 0.4 m/s, Sliding velocity: 0.008 m/s, Slippage ratio: 2%.

3. Centerless Grinding Systems and the Characteristic Equation

Since the three stability criteria influence each other and are significantly affected by the setup conditions, it is necessary to assess these stabilities as a total system—including the machine dynamic characteristics, the centerless grinding mechanism and the grinding processes. Figure 3 shows a block diagram of the centerless grinding system. The system consists of the regenerative centering mechanism [16], the regenerative function [17], the relationship between depth-of-cut and the normal grinding force, the contact stiffness of the wheels, the wheel filtering functions, and the machine dynamics [12]. The dynamic behavior of the rounding mechanism can be investigated based on the characteristic equation of the closed loop centerless grinding system in Figure 3.

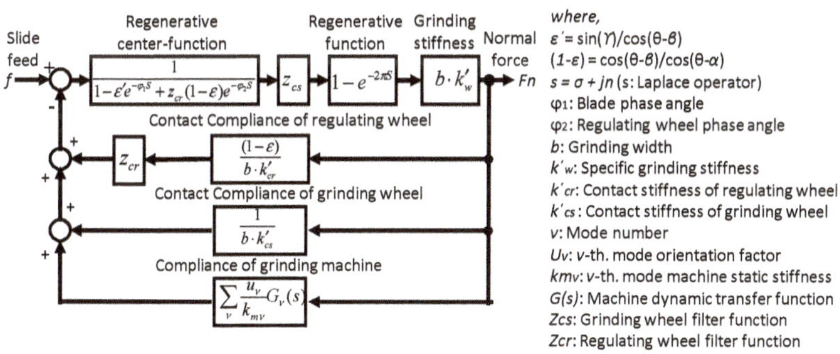

Figure 3. Block diagram of centerless grinding system.

The characteristic function is represented by:

$$-\frac{1}{f(s)} = g(s) \tag{6}$$

where

$$-\frac{1}{f(s)} = -\frac{Z_{cs}\left(1 - e^{-2\pi s}\right)}{1 - \varepsilon' e^{-\varphi_1 s} + Z_{cr}(1-\varepsilon)e^{-\varphi_2 s}} \tag{7}$$

$$g(s) = bk_w{}'\left[\frac{1}{bk'_{cs}} + \frac{(1-\varepsilon)}{bk'_{cr}} + \frac{1}{k_m}G_m(s)\right] \tag{8}$$

By solving the characteristic roots of Equation (6), the dynamic rounding stability can be evaluated and the transient behavior of the waviness amplitude in work roundness can be calculated during the grinding process. The characteristic root can be represented by:

$$s = \sigma + jn \tag{9}$$

where s is the Laplace operator, σ is the amplitude growth rate per unit radian, and n is the number of lobes in the work roundness. The transient of the amplitude change $A(t)$ on roundness waviness during the grinding process can be expressed by:

$$A(t) = A_0 \exp(2\pi n_w \sigma t) \tag{10}$$

where A_0 is the initial amplitude of the waviness, n_w is the work rotation speed in *rps* and t is the grinding time. When σ is positive, the amplitude of n lobes grows with grinding time t and the grinding process can be identified as the chatter vibration. In case of $\sigma < 0$, the amplitude of n lobes is decreased with grinding time t, and the grinding process becomes stable with improved roundness.

When the effect of machine vibration is negligible, the response of the transfer function $Gm(s)$ is degenerated to a constant and the resulting system is of a kinematic nature, referred to as "geometric rounding stability" [12]. Then, the characteristic equation is simplified as:

$$-\frac{(1-e^{-2\pi s})}{1-\varepsilon' e^{-\varphi_1 s} + (1-\varepsilon)e^{-\varphi_2 s}} = k_w{}'\left[\frac{1}{k'_{cs}} + \frac{(1-\varepsilon)}{k'_{cr}} + \frac{b}{k_m}\right] \tag{11}$$

4. Three Stability Criteria in Centerless Grinding

4.1. Work Rotation Stability Criterion

Figure 4 shows the torques acing on the workpiece during the centerless grinding process. Tg is the grinding torque given by the tangential grinding force Ft. Tb and Tr are the friction torques acting on the blade and the regulating wheel, respectively. Under the stable grinding process, the following torque-quilibrium relationship is maintained. The work peripheral velocity Vw is controlled by the friction drive/brake mechanism of RW, and becomes almost the same as the *RW* peripheral velocity Vr.

$$\text{Stable grinding}: \ Tg = Tb + Tr, Vw \cong Vr \tag{12}$$

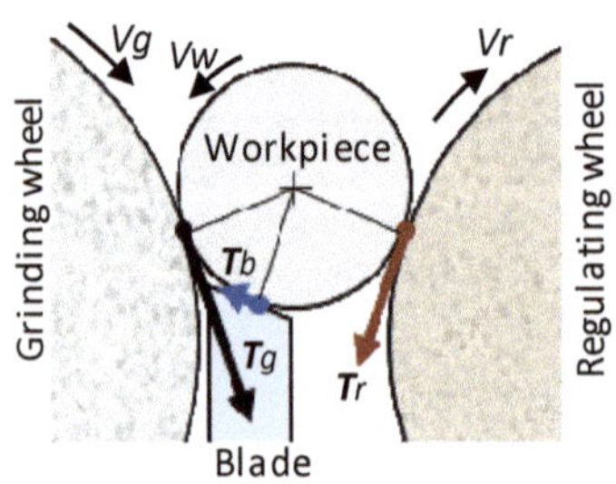

Figure 4. Torques acting on workpiece during centerless grinding.

However, once this quilibrium condition is broken by the excessive grinding torque Tg overcoming the friction torques (Tb + Tr) during grinding, the work velocity Vw suddenly increases toward the grinding wheel speed Vg.

$$\text{Unstable grinding}: \ Tg > Tb + Tr, Vw \cong Vg \tag{13}$$

This phenomenon, called "spinners", can create a potentially very dangerous situation and should be avoided for safe operations.

Figure 5a,b show the geometrical arrangement of the centerless grinding process and the forces acting on the workpiece at any cut section perpendicular to the work axis l during grinding. The variables f_T and f_N represent the tangential and normal grinding forces per unit width at the cut section. Rb and Rr are the resultant forces, while μ_b and μ_r are the friction coefficients at the contact points with the blade and the RW, respectively. *w(l)* is the work weight per unit width at the cut Section 1. The torque equilibrium equation can be written by:

$$I\frac{d\omega}{dt} = \int_0^L r(l)\frac{(B_1\mu_r(l) + B_2)f_T(l) - (C_1\mu_r(l) + C_2)w(l)}{(A_1\mu_r(l) + A_2)}dl \tag{14}$$

where

$$A_1 = \mu_b \cos(\theta - \alpha) - \sin(\theta - \alpha) \tag{15}$$

$$A_2 = \mu_b \sin(\theta - \alpha) + \cos(\theta - \alpha) \tag{16}$$

$$B_1 = A_1 - \mu_b(\sin\gamma + k\cos\gamma) - [(1 + k\mu_b)\sin(\theta + \beta) + (k - \mu_b)\cos(\theta + \beta)] \tag{17}$$

$$B_2 = A_2 - \mu_b(\cos\gamma - k\sin\gamma) \tag{18}$$

$$C_1 = \sin\theta - \mu_b(\cos\theta - \sin\alpha) \tag{19}$$

$$C_2 = \mu_b \cos\alpha \tag{20}$$

I and ω are the mass moment of inertia and the angular velocity of the workpiece. k is the force ratio (f_N/f_T). For convenience, the plus sign of μr is assigned to the downward friction force and the minus sign is assigned to the upward one.

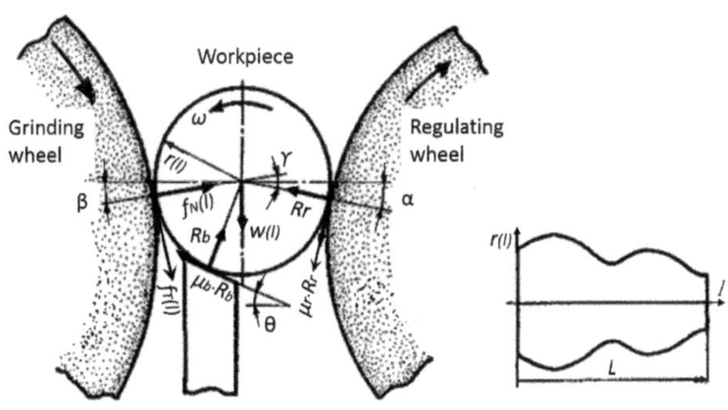

Figure 5. Arrangement of centerless grinding and forces acting on workpiece. (**a**) Forces acting on workpiece during centerless grinding; (**b**) Cylindrical workpiece.

The generalized motion Equation (14) is applicable to any cylindrical-shaped workpiece; for example, simple cylindrical, tapered, and multiple stepped diameter workpieces. Equation (14) indicates that, in addition to being affected by the primary setup conditions (θ, γ), the rotational motion of the workpiece is affected by the grinding forces and the friction force on RW.

The upper-limit tangential grinding force f_U under the stable grinding condition is derived from Equation (14).

$$f_U = \frac{(C_1\mu_{r0} + C_2)}{(B_1\mu_{r0} + B_2)}w \tag{21}$$

where μ_{r0} is the maximum static friction coefficient of RW. When the tangential grinding force f_T is smaller than f_U, the work rotation speed Vw can be controlled with the *RW* speed *Vr*.

Figure 6 shows the results of the calculation of f_U with respect to the blade angle θ with various friction coefficients μ_{r0}. The grinding force f_U is normalized with the diameter d of a simple cylindrical workpiece made of steel. f_U increases with increased θ. When θ is greater than a certain angle with μ_{r0}, the f_U value becomes infinite. Under this condition, there is no risk of the spinners phenomenon occurring. The zone with the infinite f_U value is called the "safe operation zone." For instance, there is no limit on f_U when a blade of $\theta > 42°$ is used with an *RW* of $\mu_{r0} = 0.25$.

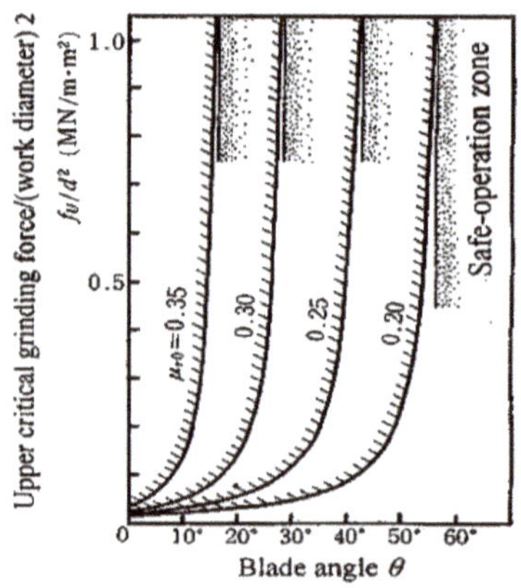

α=4.4°, β=2.6°, γ=7.0°, k=2.0, μ_{b0}=0.15

Figure 6. Upper limit tangential grinding force.

Figure 7 shows the safe operation zones under various μ_{r0} values on the (θ–γ) chart and provides guidelines for satisfying the work rotation stability criterion (WRSC). Stable grinding without any risk of spinners can be obtained by selecting the set of (θ, γ) from the safe operation zone, and the WRSC is satisfied with the setup conditions (θ, γ).

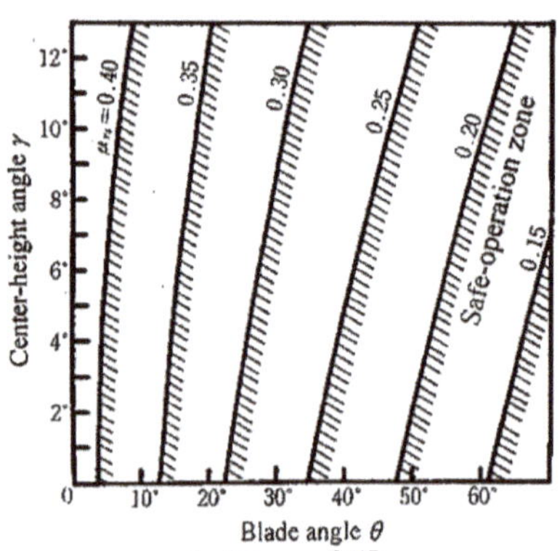

k=2.0, μ_b=0.15

Figure 7. Safe operation chart.

4.2. Geometrical Rounding Stability Criterion

When the grinding system is stable and the influence of the machine dynamics on the rounding mechanism is negligible, the stability of the rounding mechanism is predominantly affected by the geometrical arrangement of the centerless grinding system under the solid-body machine structure. By analyzing the characteristic roots of Equation (11), the effect of the center height angle γ on geometrical rounding stability can be assessed. Figure 8a shows the characteristic root distributions for the odd lobes and the even lobes [14]. When a lower center height angle such as $\gamma < 3°$ is set up, the amplitude growth rates of the 3, 5, and 7 lobes become close to zero, and the roundness error due to these odd lobes cannot be improved during the grinding process. On the other hand, when a higher center height angle such as $\gamma > 9°$ is set up, the amplitude growth rates of even lobes like 18, 20, and 22 become close to zero and the roundness error cannot be improved due to the even-lobe waviness.

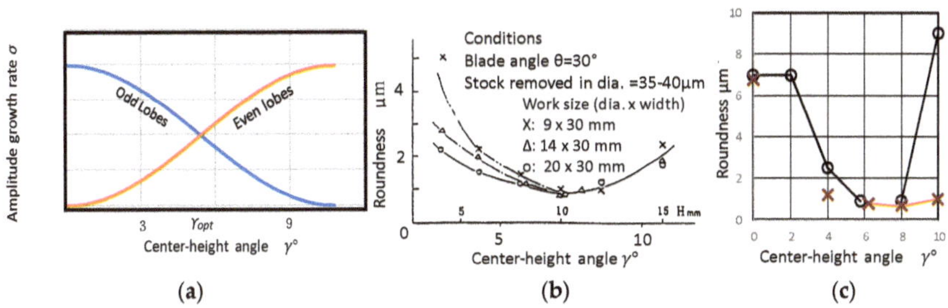

Figure 8. Geometrical rounding stability. (**a**) Center-height angle vs. amplitude growth rates [14]; (**b**) Effect of γ on roundness [6]; (**c**) Effect of γ on roundness [16].

These results suggest the existence of an optimum center height angle γopt that yields a minimum roundness error, and γopt = 6.7° has been proposed as that optimum angle [12]. Miyashita et al. reported the experimental results shown in Figure 8b, and indicated that the optimum center height angle exists around 7° [6]. Rowe et al. reported on theoretical and experimental analysis of the rounding mechanism of a workpiece with a flat face. Figure 8c shows the effect of the center height angle on the roundness error. The grinding test results verified that the optimum center height angle γopt exists around 6°–8° [8].

It is well known that an odd number of lobes appears under a lower center height condition such as $\gamma < 3°$, as shown in Figure 9a. To minimize the roundness error with odd numbers of lobes, it is recommended that lower center height angles be avoided as much as possible. Under a relatively higher center height condition, a specific even number of lobes appears, as shown in Figure 9b. Where the center height angle is known, the even number of lobes n_e that will appear can be found by $180/\gamma$ [6].

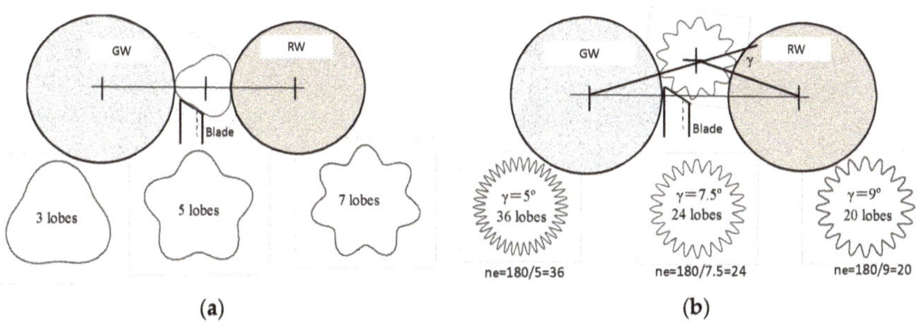

Figure 9. Effect of center-height angle on rounding mechanism. (**a**) Odd lobe appearance at lower γ; (**b**) Even lobe appearance at a specific γ.

These even numbers of lobes are the result of the geometrical rounding instability caused by the geometrical arrangement of the regulating wheel or the blade. Figures 10 and 11 show the geometrical rounding stability criteria of the regulating wheel and the blade, respectively.

Figure 10a is an example of regulating wheel geometrical rounding instability. When a peak of waviness on the work roundness contacts with the regulating wheel and the waviness becomes a valley at the grinding point (and vice versa), the waviness error cannot be removed during the grinding process. Conversely, Figure 10b is an example of regulating wheel geometrical stability. When a peak contacts with the regulating wheel and the waviness becomes a peak at the grinding point (and vice versa), the waviness error can be removed during the process. Therefore, the regulating wheel geometrical rounding stability criterion ($RW - GRSC$) can be summarized as follows:

$$(RW - GRSC)\ Unstable : \frac{180}{\gamma} = Even \text{ integer} \tag{22}$$

$$(RW - GRSC)\ Stable : \frac{180}{\gamma} = Odd \text{ integer} \tag{23}$$

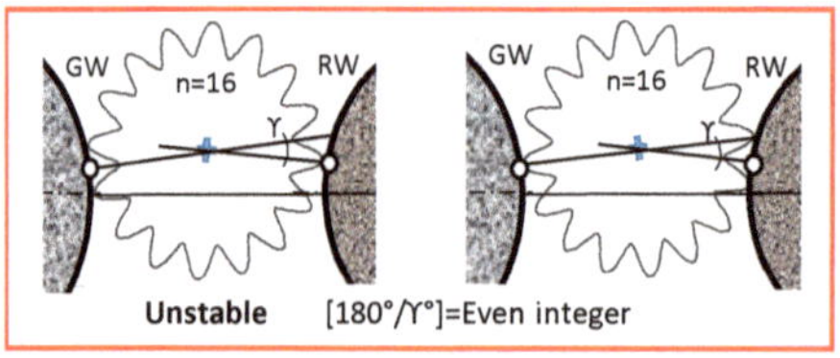

(**a**)

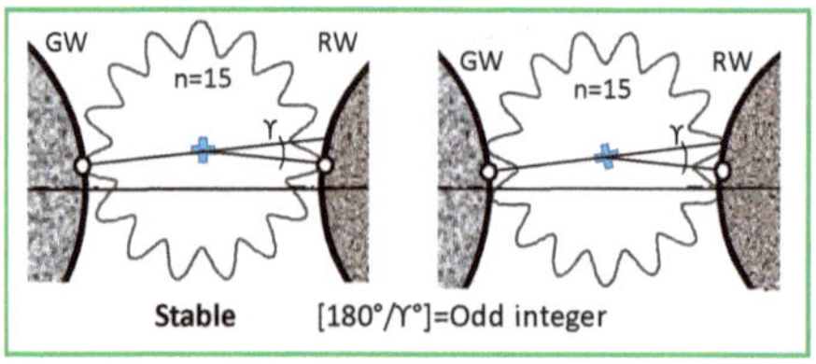

(**b**)

Figure 10. Regulating wheel (RW) geometrical rounding stability criteria. (**a**) *RW* geometrical rounding unstable conditions; (**b**) *RW* geometrical rounding stable conditions.

Likewise, Figure 11a is an example of blade geometrical rounding instability. When a peak of waviness contacts with the top surface of the blade and the waviness becomes a peak at the grinding point (and vice versa), the waviness error cannot be removed. In addition, again, Figure 11b is an example of blade geometrical rounding stability. When a valley contacts with the top surface of the blade and the waviness becomes a peak at the grinding point (and vice versa), the waviness error can be removed. Similarly, the blade geometrical rounding stability criterion ($B - GRSC$) can be summarized as follows:

$$(B - GRSC)\ Unstable : \frac{90 - \theta - \beta}{\gamma} = Even \text{ integer} \tag{24}$$

$$(B - GRSC)\ Stable : \frac{90 - \theta - \beta}{\gamma} = Odd \text{ integer} \tag{25}$$

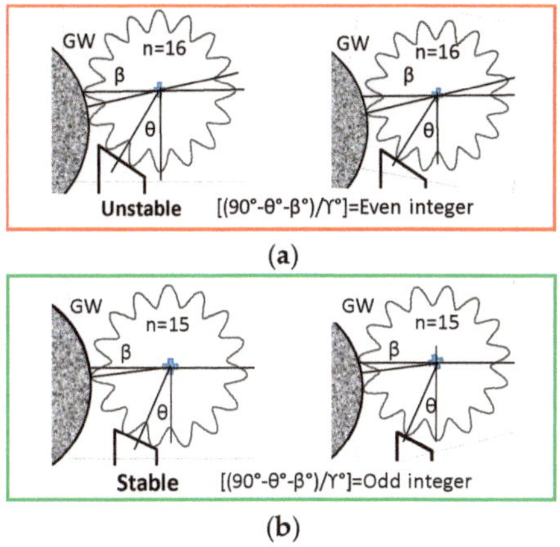

Figure 11. Blade geometrical rounding stability criteria. (**a**) Blade geometrical rounding unstable conditions; (**b**) Blade geometrical rounding stable conditions.

To satisfy the *RW* geometrical rounding stability criterion ($RW - GRSC$), the center height angle γ is determined by:

$$RW - GRSC\ Stable : \gamma = \frac{180}{I_{odd}} \tag{26}$$

where *Iodd* is an odd integer. Also, the blade geometrical stability criterion ($B - GRSC$) can be satisfied by setting up the center height angle γ that can be calculated by the following equation.

$$B - GRSC\ Stable : \gamma = \frac{(90 - \theta)}{\left[Iodd + \frac{(Dr+Dw)}{(Dg+Dr+2Dw)}\right]} \tag{27}$$

4.3. Dynamic System Stability Criterion

As mentioned, the chatter vibration in the centerless grinding system is very severe and builds up fast because the work-regenerative self-excited vibration has such a high amplitude growth rate, as shown in Figure 12a. This phenomenon not only deteriorates grinding accuracy and productivity, but also threatens safe operations. To achieve a stable grinding process with high grinding accuracy as shown in Figure 12b, satisfying the dynamic system stability criterion is imperative.

Study of the characteristic root distributions of Equation (6) clarifies the generation mechanism of the chatter vibration [12], and the chatter generation zones are revealed on the ($n{\cdot}\gamma$–$n{\cdot}n_w$) diagram for the dynamic system stability criterion (DSSC). Figure 13 plots the 3D positive growth rates σ of the characteristic roots on the ($n{\cdot}\gamma$–$n{\cdot}n_w$) diagram. The chatter zones are shown as "mountains" located near the natural frequencies in the ($n{\cdot}n_w$) axis. The higher the height of the mountain, the more severe the chatter vibration. Since the chatter mountains are in $0 < (n{\cdot}\gamma) < 180°$, they generate even numbers of lobes during chatter vibration.

The chatter zones located in $180° < (n{\cdot}\gamma) < 360°$ generate odd numbers of lobes and generate chatter vibration only under higher γ values. A straight line (n_w/γ) through the origin is determined when the center height angle γ and the work speed n_w are given. Where the straight line hits a mountain, chatter vibration occurs at that frequency. In other words, where the straight line given by the ratio (n_w/γ) does not hit any mountains, the DSSC is satisfied and chatter-free grinding is achieved. Figure 13 shows distinct areas that satisfy the DSSC and gives the ranges of (n_w/γ) that provide chatter-free grinding.

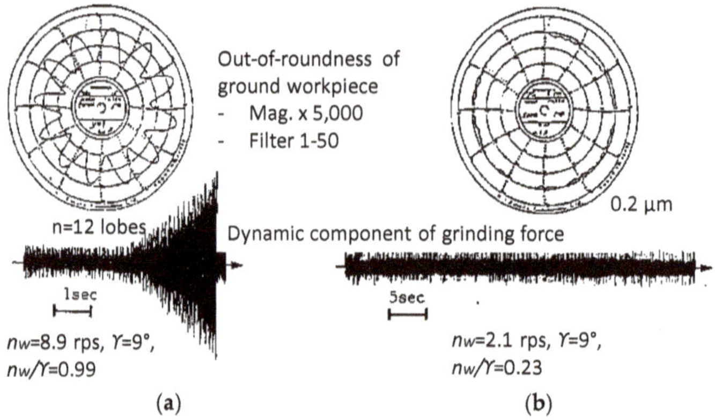

Figure 12. Dynamic stability of centerless grinding system (experiment). (**a**) Unstable grinding process (chatter); (**b**) Stable grinding process.

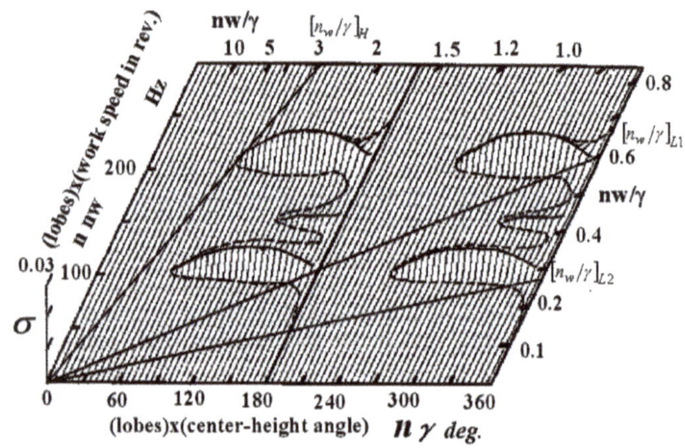

Figure 13. Dynamic system stability diagram for Machine A. Conditions: b = 70 mm, $k'w$ = 2 kN/mm·mm, $k'cr$ = 0.3 kN/mm·mm, $k'cs$ = 1 kN/mm·mm, kmr = 0.1 kN/µm, kms = 0.15 kN/µm, $km1$ = 0.3 kN/µm, fnr = 100 Hz, fns = 200 Hz, $fn1$ = 150 Hz, ζr = 0.05, ζs = 0.05, $\zeta 1$ = 0.05.

Each grinding machine has its own dynamic characteristics with different natural frequencies, and a ($n{\cdot}\gamma$–$n{\cdot}n_w$) diagram can be plotted that shows its unique chatter zones. The chatter zones are identified by conducting systematic grinding tests. Figure 14a shows the chatter zones of grinding machine A. The chatter zones located near the natural frequencies of 100 Hz, 150 Hz and 200 Hz are shown on the vertical axis ($n{\cdot}n_w$) Hz. These chatter zones are located in $0 < (n{\cdot}\gamma) < 180°$ and generate even numbers of lobes, while the zones located in $180° < (n{\cdot}\gamma) < 360°$ generate odd numbers of lobes where $\gamma > 6.7°$. Figure 14b shows the chatter zones for grinding machine B. Machine B is designed with high, rigid structures and possesses spindles with very high stiffness. The first chatter zone appears at the natural frequency of 430 Hz, which is very high, and machine B creates wider chatter-free regions than conventional machine A.

Figure 15a shows the chatter zones of grinding machine A plotted on a (γ–n_w) chart. The practical (γ–n_w) chart can describe chatter zones, but cannot describe the chatter generation mechanism, the chatter zones' various vibration modes, or the areas where stable grinding can occur. However, the (γ–n_w) chart is very effective in setup operations when used along with the analytical ($n\cdot\gamma$–$n\cdot n_w$) diagram. The ranges of chatter zones in (n_w/γ) are explicitly given, as shown in Figure 15b.

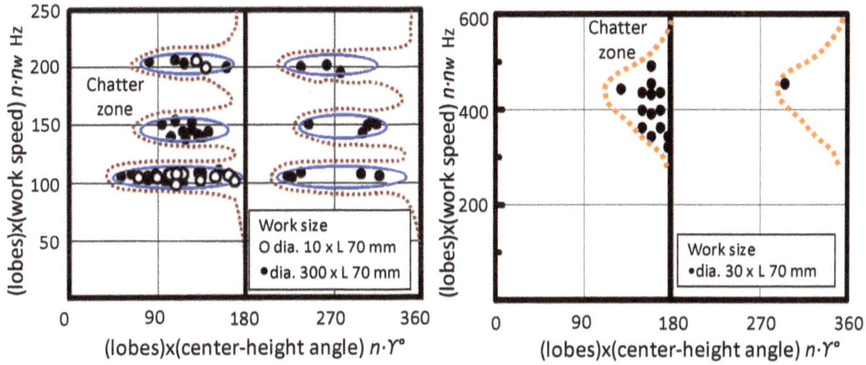

(**a**) (**b**)

Figure 14. Chatter zones on ($n\cdot\gamma$–$n\cdot n_w$) diagram. (**a**) Grinding machine A; (**b**) Grinding machine B.

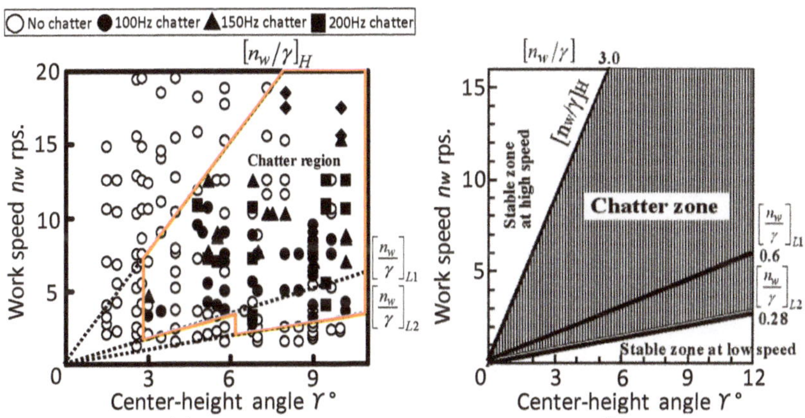

(**a**) (**b**)

Figure 15. Dynamic system stability criterion for machine A. (**a**) Chatter zones confirmed by grinding tests; (**b**) Dynamic system stability criterion.

Three types of stable, no-chatter zones exist for high work speed and low work speed regions. The chatter-free conditions for machine A are:

1. (n_w/γ) H > 3.0 (high-speed chatter-free zone; KH)
2. (n_w/γ) L1 < 0.6 when γ is lower (low-speed chatter-free zone 1; KL1)
3. (n_w/γ) L2 < 0.28 when γ is higher (low-speed chatter-free zone 2; KL2)

Figure 16a,b redraw Figure 14b to include chatter zone boundary lines. The chatter-free zones are:

4. (n_w/γ) H > 4.24 (KH)
5. (n_w/γ) L1 < 2.15 (KL1) for $\gamma < 6.67°$
6. (n_w/γ) L2 < 1.08 (KL2) for $\gamma > 6.67°$

Figure 16a shows a narrow stable zone in 1.53 < (n_w/γ) < 2.15. The setup for this chatter-stable zone is too risky to use, so it is not considered an area of stable, chatter-free conditions.

Since the work speed n_w is controlled by the regulating wheel speed Nr, it is helpful to convert the (γ–n_w) charts (Figures 15b and 16b) into a (γ–N_r) chart, as shown in Figure 17. For practical setup operations, the range of the center height angle is set to γ = 3° to 9°. Also, in this case the range of the speed ratio q (defined as the ratio of the work speed Vw to the grinding speed Vg) for surface roughness control is set to $1/q$ = 25 to 150. In Figure 17, three chatter-free stable zones—KH, KL1 and KL2—are shown within the constrained range. The dynamic system stability criterion is satisfied when a set of (γ, Nr) is selected from the chatter-free stable zones KH, KL1 and KL2.

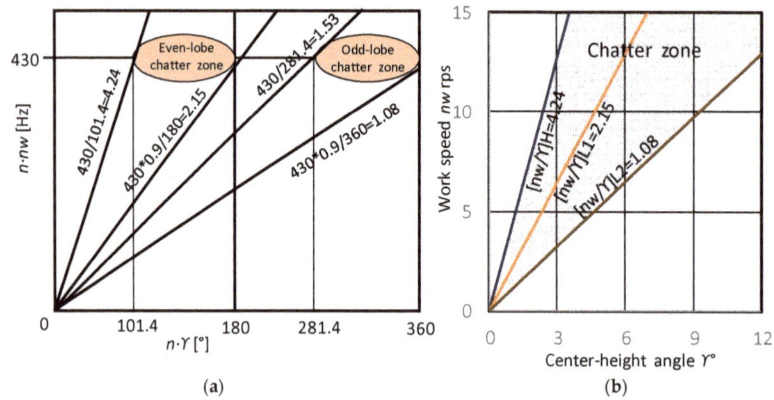

Figure 16. Dynamic system stability criterion for machine B. (**a**) Chatter zones on ($n{\cdot}\gamma$–$n{\cdot}n_w$) diagram; (**b**) Chatter zones on (γ–n_w) chart.

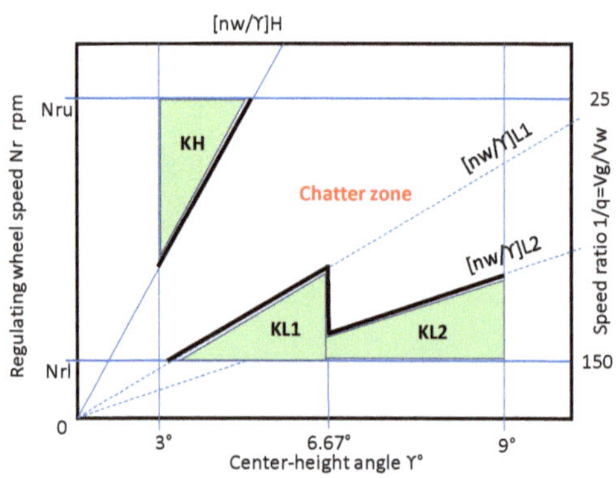

Figure 17. Chatter free zones on γ–Nr chart. KH: High speed chatter-free zone, KL1, KL2: Low speed chatter-free zones.

5. Modeling to Find the Optimum Setup Conditions that Satisfy the Three Stability Criteria of Centerless Grinding

The previous section discussed the determination of setup conditions that would satisfy each individual stability criterion. This section discusses the development of a model for the optimum setup conditions that will simultaneously satisfy all three stability criteria.

Figure 18 describes the structure of the developed model. As the first step, the sets of (θ, γ) (blade angle θ, center-height angle γ) satisfying the three individual stability criteria are determined. To satisfy the work rotation stability criterion, the sets of (θ, γ) are selected from the safe operation zone shown in Figure 7, analyzed with the varying maximum friction coefficient μ_{r0} of a given regulating wheel. Also, the sets of (θ, γ) satisfying both geometrical rounding stability criteria—$RW - GRSC$ and $B - GRSC$—are calculated. Then, the sets of (n_w, γ) (work speed n_w and γ) that satisfy the dynamic system stability criterion are found. The sets of (n_w, γ) are selected from one of the stable chatter-free zones: KH, KL1 or KL2 (see Figure 17).

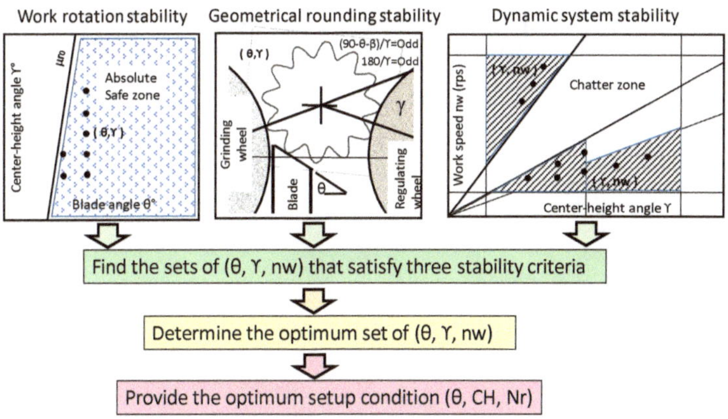

Figure 18. Structure of the model for determining the optimum setup condition

The second step is to find the sets of (θ, γ, n_w) satisfying all three stability criteria. The third step is to determine the optimum set from among the population of the (θ, γ, n_w) sets by calculating the PI (performance index) function based on the process aim (accuracy or productivity). Finally, the optimum set (θ, γ, n_w) is converted into practical setup conditions (blade angle θ, center height CH, RW speed Nr) as the outputs of the developed model.

Figure 19 is the flow chart of "Opt-Setup Master", the developed model. The model requires a machine operator to provide some input information, as shown in Table 1. All parameters required for the calculation of the Opt-Setup Master are referenced from the data bank, which stores machine specifications, machine dynamic characteristics, work part numbers with dimensions, *RW* friction characteristics, blade availability, etc. The constraints of the setup parameters are also stored in the data bank, as shown in Table 2.

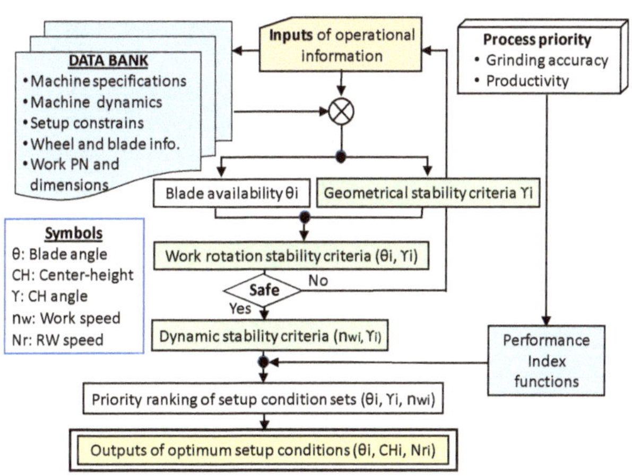

Figure 19. Flow chart of developed model "Opt-Setup Master".

Table 1. Input information and parameters referred to data bank.

Inputs	Action	Parameters Referred to Data Bank
Machine name	Select	Machine specifications, Machine dynamic characteristics (Natural frequencies, damping ratios)
Workpiece shape	Select	Cylindrical (CYD), Tapered (TPD), Spherical (SRL), Multi-stepped (STD)
Workpiece part-number	Select	Dimensions (diameter, length, etc.), profile
GW diameter	Measure	New, worn, measured
RW diameter	Measure	New, worn, measured
RW dresser type and dress lead	Select	Single point dress, Rotary dress, Friction coefficient of RW
Blade availability	Select	Blade angle θ, blade thickness t

Table 2. Constraints of setup parameters.

Setup Parameters	Symbol	Unit.	Min.	Max.	Typical
Range of speed ratio	$1/q = Vg/Vw$	-	25	150	100
Range of blade angle	θ	°	15	45	30
Range of Center-height angle	γ	°	3	9	6.67
Range of regulating wheel speed	Nr	rpm	15	100	50
Range of *GW* diameter	Dg	mm	375	455	450
Range of *RW* diameter	Dr	mm	275	350	345
Range of Workpiece diameter	Dw	mm	5	100	40
Grinding wheel speed in revolution	Ng	rpm	1260	2300	1890
Grinding speed	Vg	m/s	30	45	45

The Opt Setup Master calculates the boundary line of the safe operation zone on the safe operation chart by using the following relationship between the dressing lead $lead_r$ and the maximum friction coefficient μ_{r0} of the rubber bonded regulating wheel [15].

$$\mu_{r0} = a \cdot lead_r + 0.33 \tag{28}$$

where a is a constant. a = 0.14 and a = 0.02 for a SPD (single-point dresser) and an RD (rotary dresser), respectively.

The boundary line of the safe operation zone on the (θ, γ) chart can be expressed by:

$$\gamma_c = m \cdot \theta_c + b \tag{29}$$

where

$$m = 3.036 \cdot a \cdot lead_r + 1.168 \tag{30}$$

$$b = 77.06 \cdot (a \cdot lead_r)^2 - 27.55 \cdot (a \cdot lead_r) - 21.23 \tag{31}$$

When $\theta 1$ of a point $(\theta 1, \gamma 1)$ is greater than $(\gamma 1 - b)/m$, the point is located at the right side of the safe operation zone boundary line and meets the work rotation stability criterion.

The first calculation of the Opt-Setup Master is to find the sets of $(\theta g, \gamma g)$ that meet the geometrical rounding stability criteria $RW - GRSC$ and $B - GRSC$. The next calculation is to determine if the sets $(\theta g, \gamma g)$ are located within the safe operation zone under the given dressing conditions of the rubber bonded regulating wheel. If they are (answer "yes"), the sets $(\theta gw, \gamma gw)$ satisfy both the geometrical rounding and the work rotation stability criteria. The final calculation is to find the work speed n_w by using γgw and the chatter stability boundary lines of $(n_w/\gamma gw)$. From these calculations, the optimum sets of (θ, γ, n_w) are discovered.

Then, the performance index (PI) functions that were prepared based on the process aims are applied to assess the optimum sets of (θ, γ, n_w). The PI functions are summarized in Table 3. The weighting factors of the PI functions are determined by applying theoretical knowledge, experimental knowledge and operational skills. PI functions can be updated with newly gained knowledge and skills. For each process aim (accuracy or productivity), the values of the PI function for all setup conditions are calculated and these sets are ranked in ascending order from minimum to those with greater values. The smaller the PI values, the more suitable the setup conditions. The setup conditions with the smallest PI function values are defined as the optimum setup conditions.

Table 3. Performance Index (PI) functions.

Process Aim	PI Function	Weighting Factors				
		A	B	C	D	E
Accuracy	$PIa = A \times I\theta - 27.5I + B \times I\gamma - 6.67I + C \times INr - 50I + D \times FL + E \times STYP$	0.2	2	0.1	0.01	2
Productivity	$PIr = A \times I\theta - 45I + B \times I\gamma - 5.15I + C \times INr - 50I + D \times FL + E \times STYP$	0.1	2	0.12	0.5	2

θ: Blade angle; γ: Center-height angle; Nr: RW speed in rpm; FL: KL1 = 0, KL2 = 0.5, KH = 1, STYP: $(RW + B)$ $GRSC = -1$, $RW - GRSC = 0$, $B - GRSC = +1$.

Table 4 shows examples of the Opt-Setup Master outputs. The conditions of the model simulation are:

(1) machine B is applied
(2) the process aim is accuracy-oriented
(3) the work shape and the size are a cylindrical workpiece with 40 mm (D) × 60 mm (L)
(4) the rubber bonded RW is dressed with $lead_r$ = 0.5 mm/rev by SPD
(5) the GW diameter is 453 mm
(6) the RW diameter is 350 mm
(7) the available blade angles are $\theta = 27.5°$ and $\theta = 40.3°$

Table 4. Examples of outputs from Opt-Setup Master.

Priority	Optimum Set Up Conditions				Engineering Parameter				Stability Parameters		
Ranking	Blade Angle	Blade Thickness	Center-Height	*RW* Speed	CH Angle	Work Speed	1/q Ratio	(n_w/γ)	Stable Zone	*RW* − *GRSC*	*Blade* − *GRSC*
No.	θ (°)	t (mm)	CH (mm)	Nr (rpm)	γ (°)	n_w (rps)	Vg/Vw	(1/s)	KH/KL1/KL2	$180/\gamma$	$(90-\theta-\beta)/\gamma$
1	40.3	20	12.68	41.8	6.68	6.1	25	0.91	KL2	27	7
2	40.3	20	12.66	83.5	6.67	12.2	25	1.83	KL1	27	7
3	40.3	20	13.67	45.1	7.2	6.58	25	0.91	KL2	27	6.5

Conditions: Machine B, Process aim: accuracy, Work: cylindrical Dia.40 × L60 mm.

Ranking No. 1 has the optimum setup conditions, as it has the smallest PI values. The practical setup parameters are a blade angle $\theta = 40.3°$ with a thickness of 20 mm, a center height of CH = 12.68 mm and RW speed of Nr = 41.8 rpm. The center height angle is $\gamma = 6.68°$, the work speed is n_w = 6.1 rps and the $1/q$ is 25.

The optimum setup condition set was selected from the safe operating zone. Therefore, it meets the work rotation stability criterion and ensures safe operations. Further, the optimum setup condition set was selected from the stable zone KL2 for chatter-free grinding, and thus meets the dynamic system stability criterion. Also, the optimum setup condition set meets the criteria of both Equations (23) and (25) (180/6.68 = 27 and (90 − 40.3 − 2.95)/6.68 = 7), so the geometrical rounding stability criteria are maintained.

Through these means, it is verified that the optimum setup condition set—provided as the outputs from the Opt-Setup Master—simultaneously satisfies all three stability criteria for centerless grinding.

Table 5 shows the optimum setup conditions as calculated by the Opt-Setup Master for infeed centerless grinding of cylindrical workpieces of various sizes by two different grinding machines, A and B. In all cases, $\gamma = 6.67°$, one of the most preferable γ angles, is chosen. All values representing $RW - GRSC$ and $B - GRSC$ are odd integers, indicating that all the setup conditions meet the geometrical rounding stability criteria. Machine B has a greater chatter DSSC index (n_w/γ) than machine A in chatter-stable zones KH, KL1 and KL2. Machine B's high stiffness creates more extensive chatter-stable zones than conventional machine A.

Table 5. The optimum setup conditions provided by the Opt-Setup Master for the grinding of cylindrical workpieces with different machines A and B.

Case No.	Work Dia.x L.	Machine	Blade Angle	C-H Angle	*RW* Speed	Chatter DSSC	Chatter Stable Zone	$RW - GRSC$	$Blade - GRSC$
	mm	A/B	θ (°)	γ (°)	Nr (rpm)	(n_w/γ)	KH/KL1/KL2	$180/\gamma$	$(90 - \theta - \beta)/\gamma$
1	10 × 20	A	27.5	6.67	38.6	3.37	KH	27	9
2		B	27.5	6.67	64.1	5.61	KH	27	9
3	20 × 30	A	27.5	6.67	77.1	3.37	KH	27	9
4		B	27.5	6.67	41.8	1.83	KL1	27	9
5	30 × 50	A	40.3	6.67	14.6	0.43	KL1	27	7
6		B	40.3	6.67	62.7	1.83	KL1	27	7
7	40 × 60	A	40.30	6.67	19.4	0.43	KL1	27	7
8		B	40.30	6.68	41.8	0.91	KL2	27	7
9	50 × 70	A	40.30	6.67	24.3	0.43	KL1	27	7
10		B	40.30	6.68	52.3	0.91	KL2	27	7
11	60 × 80	A	40.30	6.67	29.1	0.43	KL1	27	7
12		B	40.30	6.68	62.7	0.91	KL2	27	7

Condition: *GW* ϕ453 mm, Rubber bonded *RW* ϕ350 mm, $\mu_{r0} = 0.4$, Available blade $\theta = 27.5°, 40.3°$.

6. Conclusions

Centerless grinding systems possess some unique features, including their work rotation drive, loose work holding and regenerative centering mechanisms. However, because of these features, three fundamental stability issues arise. Many researchers have investigated the issues and provided useful guidelines for solving the issues.

This paper summarizes the three fundamental stability issues: (1) work rotation stability for safe operation with no spinners; (2) geometrical rounding stability for better roundness; and (3) dynamic system stability for chatter-free grinding. It emphasizes the need for an analytical tool that can provide optimal setup conditions—those conditions that will satisfy all three stability criteria simultaneously. This paper describes a newly developed analytical tool named Opt-Setup Master, and discusses how the three stability criteria can be met.

The developed Opt-Setup Master has the following features:

(1) Accepts various shapes of workpiece: cylindrical, tapered, spherical and multi-stepped
(2) Applicable to any centerless grinding machine
(3) Data management via a data bank
(4) Inputs are easy to enter and outputs are readily usable
(5) Designed for operators
(6) Provides scientific parameters for engineers/managers
(7) Finds all setup conditions satisfying the three stability criteria of centerless grinding systems
(8) Outputs the optimum condition based on process aim

Conflicts of Interest: The authors declare no conflict of interest.

References

1. Hashimoto, F.; Gallego, I.; Oliveira, J.F.G.; Barrenetxea, D.; Takahashi, M.; Sakakibara, K.; Stålfelt, H.-O.; Staadt, G.; Ogawa, K. Advances in Centerless Grinding Technology. *CIRP Ann. Manuf. Technol.* **2012**, *61*, 747–770. [CrossRef]
2. Hashimoto, F.; Suzuki, N.; Kanai, A.; Miyashita, M. Critical Range of Set-up Conditions of Centerless Grinding and Problem of Safe Machining Operation. *Jpn. Soc. Precis. Eng.* **1982**, *48*, 996–1001. [CrossRef]
3. Hashimoto, F.; Lahoti, G.; Miyashita, M. Safe Operations and Friction Characteristics of Regulating Wheel in Centerless Grinding. *Ann. CIRP* **1998**, *47*, 281–286. [CrossRef]

4. Abrasive Industry. *Anon, Centerless Grinding Operation*; Abrasive Industry: Auckland, New Zealand, 1925; pp. 102–105.
5. Dall, A.H. Rounding Effect in Centerless Grinding. *Mech. Eng.* **1946**, *68*, 325–332.
6. Ogawa, M.; Miyashita, M. On Centerless Grinding (1)—Theoretical Treatise on Rounding Action. *Jpn. Soc. Precis. Eng.* **1957**, *24*, 89–94. [CrossRef]
7. Reeka, D. On the Relationship between the Geometry of the Grinding Gap and the Roundness Error in Centerless Grinding. Ph.D. Thesis, RWTH, Aachen University, Aachen, Germany, 1967.
8. Rowe, W.B.; Barash, M.M.; Koenigsberger, F. Some Roundness Characteristics of Centerless Grinding. *Int. J. Mach. Tool Des. Res.* **1965**, *5*, 203–215. [CrossRef]
9. Yonetsu, S. Study on Centerless Grinding—First Report. *Jpn. Soc. Mech. Eng.* **1953**, *19*, 53–59.
10. Gurney, J.P. An Analysis of Centerless Grinding. *Trans. ASME J. Eng. Ind.* **1964**, *86*, 163–174. [CrossRef]
11. Furukawa, Y.; Miyashita, M.; Shiozaki, S. Vibration Analysis and Work Rounding Mechanism in Centerless Grinding. *Int. J. Mach. Tool Des. Res.* **1970**, *44*, 145–175. [CrossRef]
12. Hashimoto, F.; Zhou, S.S.; Lahoti, G.D.; Miyashita, M. Stability Diagram for Chatter Free Centerless Grinding and Its Application in Machine Development. *Ann. CIRP* **2000**, *49*, 225–230. [CrossRef]
13. Gallego, I. Intelligent Centerless Grinding: Global Solution for Process Instabilities and Optimal Cycle Design. *Ann. CIRP* **2007**, *56*, 347–352. [CrossRef]
14. Hashimoto, F.; Lahoti, G.D. Optimization of Set-up Conditions for Stability of the Centerless Grinding Process. *Ann. CIRP* **2004**, *53*, 271–274. [CrossRef]
15. Hashimoto, F. Effects of Friction and Wear Characteristics of Regulating Wheel on Centerless Grinding. In Proceedings of the SME, Third International Machining and Grinding Conference, Cincinnati, OH, USA, 4–7 October 1999; pp. 1–18.
16. Rowe, W.B.; Koenigsberger, F. The Work-Regenerative Effect in Centerless Grinding. *Int. J. Mach. Tool Des. Res.* **1964**, *4*, 175–187. [CrossRef]
17. Snoeys, R.; Brown, D. Dominating Parameters in Grinding Wheel and Workpiece Regenerative Chatter. In Proceedings of the 10th International Machine Tool Design and Research Conference, LanPleacaster, UK, 10–14 September 1969; Pergamon Press: Oxford, UK, 1970; pp. 325–348.

MDPI
St. Alban-Anlage 66
4052 Basel
Switzerland
Tel. +41 61 683 77 34
Fax +41 61 302 89 18
www.mdpi.com

Inventions Editorial Office
E-mail: inventions@mdpi.com
www.mdpi.com/journal/inventions

www.ingramcontent.com/pod-product-compliance
Lightning Source LLC
Chambersburg PA
CBHW041755220926
51847CB00014B/213